Structural Analysis

The Analytical Method

Structural Analysis

The Analytical Method

Ramon V. Jarquio, P.E.

CRC Press
Taylor & Francis Group
Boca Raton London New York

CRC Press is an imprint of the
Taylor & Francis Group, an **informa** business

CRC Press
Taylor & Francis Group
6000 Broken Sound Parkway NW, Suite 300
Boca Raton, FL 33487-2742

First issued in hardback 2019

© 2008 by Taylor & Francis Group, LLC
CRC Press is an imprint of Taylor & Francis Group, an Informa business

No claim to original U.S. Government works

ISBN-13: 978-1-4200-6023-2 (hbk)

Library of Congress Cataloging-in-Publication Data
Jarquio, Ramon V.
Structural analysis : the analytical method / Ramon V. Jarquio.
p. cm.
Includes bibliographical references and index.
ISBN 978-1-4200-6023-2 (alk. paper)
1. Structural analysis (Engineering) I. Title.
TA645.J37 2007
624.1'7--dc22   2007005295

Visit the Taylor & Francis Web site at
http://www.taylorandfrancis.com

and the CRC Press Web site at
http://www.crcpress.com

Preface

This book illustrates the analytical procedures for predicting the capacities of circular and rectangular sections in concrete and steel materials. It introduces the capacity axis in the analysis, which is a geometric property not considered in all the current solutions in standard literature. It precludes the use of the current standard interaction formula for biaxial bending, which is a crude and inefficient method. More importantly, the analytical method will prove the necessity of utilizing the capacity axis not only for determining the minimum capacity of a section for biaxial bending but also as a reference axis to satisfy the equilibrium of external and internal forces. Under the current standard interaction formula for biaxial bending, the satisfaction of equilibrium conditions is not possible. Proving the equilibrium condition is the fundamental principle in structural mechanics that every analyst should be able to do.

Chapter 1 covers the derivation of equations required for the prediction of the capacity of the footing foundation subjected to a planar distribution of stress from soil bearing pressures. The capacity of the footing is defined by a curve wherein the vertical axis represents the scale for total vertical load on the footing, and the horizontal axis represents the scale for maximum moment uplift capacity. This capacity curve encompasses all states of loading in the footing including cases when part of the footing is in tension. Hence, it becomes an easier task for a structural engineer to determine whether a given footing with a known allowable soil pressure is adequate to support the external loads. There is no need to solve for biquadratic equations to determine solutions for a footing with tension on part of its area. The Excel spreadsheet only requires entering the variable parameters such as footing dimensions and allowable maximum soil pressure.

The procedures in the derivation are very useful in the prediction of capacities of steel sections that are normally subjected to linear stress conditions. This is the subject of Chapter 2.

Chapter 1 also includes the derivation of Boussinesq's elastic equation for the dispersion of uniform and triangular surface loads through the soil medium. The derived equations will be useful in the exact value of average pressure to apply in the standard interaction formula for settlement of footing foundations without using the current finite-element method or charts for this problem.

Chapter 2 deals with the application of the analytical method to predict the capacities of steel pipe in circular or rectangular sections. The steps in Chapter 1 together with the principle of superposition will be utilized to derive the equations for the square and rectangular tubular sections. Equations for the outer section are derived first, followed by the inner section. The difference between the outer and inner section will determine the yield capacity of any steel tubing.

The equations derived for the rectangular section in conjunction with the principle of superposition will be utilized for the steel I-sections. Here, the position of the capacity axis is chosen at the diagonal of the outer rectangular section. The value calculated along the capacity axis represents the component of the resultant bending moment capacity of the section. To obtain the resultant bending moment requires the calculation of the component of the resultant perpendicular to the capacity axis.

These equations are programmed in an Excel 97 worksheet to obtain the tabulated values at key points in the capacity curves of commercially produced steel tubing and I-sections listed by the *AISC Steel Manual*. The variable parameters to be entered in these worksheets are the dimensions of the section and the steel stress allowed. Tabulated numerical values are shown in English and System International (SI) units for yield stress of $f_y = 36$ ksi (248 MPa). For stresses, a direct proportion can be applied to the values shown in the tables. Checking the adequacy of a design using these tables is relatively easy and quickly performed by comparing the external loads to the capacities of the section at key points.

All cases of loading including that of Euler are within the envelope of the capacity curve. There is no need to know Euler load to develop this capacity curve. The particular Euler load is determined as an external loading. All that is needed is to plot the external loads and determine whether the selected section is adequate to support the external loads. If not, another section is tried until the external loads are within the envelope of the capacity curve.

Chapter 3 is the analytical method for predicting the capacities of reinforced concrete circular and rectangular columns using the familiar Concrete Reinforcing Steel Institute (CRSI) stress–strain diagram. This is in contrast with the modified stress-strain distribution for these columns used by the author in his first book, *Analytical Method in Reinforced Concrete.*

The variable parameters to enter in the Excel worksheets are the dimensions of the column section, the ultimate concrete compressive stress, the yield stress of reinforcing rods, the concrete cover to center of main reinforcement, and the number of main bars. This software program makes it easier for structural engineers to determine columns subjected to direct stress plus bending without knowing the Euler load beforehand.

The envelope of the column capacity chart includes all cases of loading including Euler's loading defined under the category of uncracked and cracked conditions in the concrete section. With this chart, all that is needed is for the structural engineer to determine the external loads and then plot them on the chart to determine the adequacy of the column to resist the external loads.

Practicing structural and civil engineers involved in the design and construction of concrete and steel structures will have a ready reference for checking the adequacy of their designs in reinforced concrete and steel sections. They can still use the traditional method they are accustomed to, but they will now have a reference of the potential capacity of the section they are dealing with. Professors as well as students will benefit from the analytical approach illustrated in this book.

The analytical method illustrated in this book is limited to circular and rectangular sections subjected to direct stress plus bending. For analysis of shear and torsion, the reader may refer to other books dealing with these stresses.

The dissemination of the information in this book includes papers presented by the author in several international conferences conducted by the International Structural Engineering and Construction (ISEC), Structural Engineering Mechanics and Computations (SEMC), and American Society of Civil Engineers (ASCE). The analytical method will give the civil and structural engineering profession a better tool in predicting capacities of structural sections used in the design of structures.

Acknowledgements go to the above-listed organizations for allowing the presentation of articles written by the author to describe the analytical method in structural analysis.

Ramon V. Jarquio, P.E.
rvjarquio@aol.com
Website: www.ramonjarquio.com

Table of contents

chapter one

Footing foundation

1.1 Introduction

Design of a rectangular footing to resist vertical and bending moment loads is done by trial and error (Bowles, 1979; Holtz and Kovacs, 1981). When the whole footing area is in compression, the standard flexure formula is applicable for equilibrium of external loads and footing capacity. When part of the area of the footing is in tension, the calculation of the actual area of footing in tension involves solving the resulting biquadratic equation. Then the uplift force and moment is determined from the specific compressive depth that will ensure equilibrium of the external and internal forces on the footing.

The planar distribution of uplift forces is the basis for determining the footing resistance to the external loads. To preclude the solution of the biquadratic equation, the footing capacity curve will have to be solved and values plotted that will encompass all cases of loading, including that when part of the footing is in tension. The variables are the width b and depth d of the footing, the allowable soil bearing pressure, and the inclination of the footing capacity axis θ with the horizontal axis. This axis may be assumed at the diagonal of the rectangular section or at a greater value of θ to determine the maximum footing resistance to external loads. The footing capacity curve is calculated using this axis.

The footing capacity curve for any value of θ can be calculated as, for instance, when $\theta = 0$ for uniaxial bending moment. This is the case for a retaining-wall footing foundation. The total vertical uplift force and the resultant moment uplift around the centerline of the footing can be calculated from the derived equations. The derived equations are then programmed in an Excel worksheet to obtain the capacity curve, which is a plot of the total vertical uplift and the resultant moment uplift for any rectangular footing with a given allowable soil-bearing pressure.

1.2   Derivation

1.2.1   Rectangular footing

Figure 1.1 shows the rectangular footing with width b and depth d acted upon by a triangular soil-bearing pressure q at any depth of compression c of the footing area. Draw lines through the corners of the rectangular area perpendicular to the X-axis. With these lines, divide the area of the rectangle into V_1, V_2, and V_3 zones to represent forces and V_1x_1, V_2x_2, and V_3x_3 as their corresponding bending moments around the Z-axis. In the XY plane, draw the stress diagram, which is a straight line passing through the position of the compressive depth c. Label the X-axis as the capacity axis and the Z-axis as the moment axis. Write the equations for the dimensional parameters as follows:

$$\text{Let } \alpha = \arctan\,(b/d) \tag{1.1}$$

$$h = d\,\cos\,\theta + b\,\sin\,\theta \tag{1.2}$$

$$z_0 = 0.50(b\,\cos\,\theta - d\,\sin\,\theta) \quad \text{when} \quad \theta < [(\pi/2) - \alpha] \tag{1.3}$$

$$z_0 = 0.50\,(d\,\sin\,\theta - b\,\cos\,\theta) \quad \text{when} \quad \theta > [(\pi/2) - \alpha] \tag{1.4}$$

$$x_2 = 0.50[d\,\cos\,\theta - b\,\sin\,\theta] \tag{1.5}$$

in which θ = axis of footing bearing capacity.

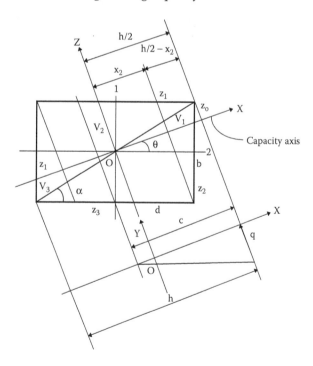

Figure 1.1 Rectangular footing foundation.

Determine the limits of the V_1, V_2, and V_3 zones to represent forces and V_1x_1, V_2x_2, and V_3x_3 as their corresponding bending moments around the Z-axis. The forces are represented by the pressure volumes whose limits are as follows: V_1 is the volume of pressure between the limit x_2 to $h/2$, V_2 is the volume of pressure from $-x_2$ to x_2, and V_3 is the volume of pressure between the limits $-(c - h/2)$ to $-x_2$.

Write the coordinates of the corner wherein line z_1 and line z_2 meet. The abscissa of this point is $h/2$ and the ordinate is z_o. Similarly, write the coordinates of the corner where line z_3 and line z_4 meet. The abscissa of this point is $-h/2$ and the ordinate is $-z_o$. From analytic geometry the point–slope formula for a straight line is of the form $y - y_1 = m(x - x_1)$. In our case $y = z$, $y_1 = z_o$, $x_1 = h/2$, and $m = -\tan\theta$. Hence, we can write the equations of the sides of the rectangular footing as follows:

$$z_1 = -\tan\theta\,(x - h/2) + z_o \tag{1.6}$$

$$z_2 = \cot\theta\,(x - h/2) + z_o \tag{1.7}$$

$$z_3 = -\tan\theta\,(x + h/2) - z_o \tag{1.8}$$

$$z_4 = \cot\theta\,(x + h/2) - z_o \tag{1.9}$$

Then write the equation of the pressure diagram using the above procedure as (Smith, Longley, and Granville, 1941):

$$y = (q/c)\{x + (c - h/2)\} \tag{1.10}$$

We are now ready to formulate the derivative of the pressure volumes using our knowledge of basic calculus. There are four cases for the envelope of values of the compressive c as follows:

Case 1: $0 < c < (h/2 - x_2)$. Three corners of the rectangular footing with negative (tension) pressures. There are two sets of limits for V_1, namely $[(h/2) - c]$ to $(h/2)$ and x_2 to $(h/2)$. The derivative for V_1 is

$$dV_1 = (z_1 - z_2)\,y\,dx$$

or, $\quad dV_1 = -(q/c)(\cot\theta + \tan\theta)\{x^2 + (c - h)x - (h/2)[c - (h/2)]\}dx \quad (1.11)$

Integrating the first set of limits to obtain

$$V_1 = -(q/6c)(\cot\theta + \tan\theta)\{2h^3/8 + 3h^2/4(c - h/2) - 3h^2/2(c - h/2)$$

$$- [2(h/2 - c)^3 + 3(h/2 - c)^2(c - h) - 3h(h/2 - c)(c - h/2)]\} \tag{1.12}$$

Simplify to

$$V_1 = (q/6)(\cot\theta + \tan\theta)c^2 \tag{1.13}$$

$$V_1x_1 = -(q/12c)(\cot\theta + \tan\theta)\{3(h^4/16) + 4(h^3/8)(c - h) + 3h(h^2/4)\,(c - h/2)$$

$$- [3(h/2 - c)^4 + 4(h/2 - c)^3(c - h) + 3(h/2 - c)^2h(c - h/2)]\} \tag{1.14}$$

Simplify to

$$V_1 x_1 = (q/12)(\cot \theta + \tan \theta)c^2(h - c) \tag{1.15}$$

Integrate and evaluate for the second set of limits for V_1 and obtain

$$V_1 = -(q/6)(\cot\theta + \tan\theta)\left\{2(h^3/8) + 3h^2/4(c-h) - 3h^2/2(c-h/2)\right.$$
$$\left. - \left[2x_2^3 + 3x_2^2(c-h) - 3hx_2(c-h/2)\right]\right\} \tag{1.16}$$

Simplify to

$$V_1 = (q/6c)(\cot\theta + \tan\theta)\left\{(3c-h)(h^2/4) + x_2\left[2x_2^2 - 3x_2(h-c) - 3h(c-h/2)\right]\right\} \tag{1.17}$$

$$V_1 x_1 = -(q/12c)(\cot\theta + \tan\theta)\left\{3(h^4/16) + 4h^3/8)(c-h) - 3h(h^2/4(c-h/2)\right.$$
$$\left. - \left[3x_2^4 + 4x_2^3(c-h) - 3hx_2^2(c-h/2)\right]\right\} \tag{1.18}$$

Simplify to

$$V_1 x_1 = (q/12c)(\cot\theta + \tan\theta)\left\{(h^3/4)(c-h/4) + x_2^2\left[3x_2^2 - 4x_2(h-c) - 3h(c-h/2)\right]\right\} \tag{1.19}$$

Case 2: $(h/2 - x_2) < c < (h/2 + x_2)$. Two corners of the rectangular footing with negative pressures. There are two limits for V_2, namely $-[c - (h/2)]$ to x_2 and $-x_2$ to x_2. The derivative for V_2 is

$$dV_2 = (z_1 - z_3)\, ydx \quad \text{or,} \quad dV_2 = (q/c)(h\tan\theta + 2z_o)\{x + [c - (h/2)]\}dx \tag{1.20}$$

Integrate the first set of limits to obtain

$$V_2 = (q/c)(h\tan\theta + 2z_o)(1/2)\{x^2 + 2[c - (h/2)]x\} \tag{1.21}$$

Evaluate limits to

$$V_2 = (q/c)(h\tan\theta + 2z_o)(1/2)\left\{x_2^2 + 2x_2(c-h/2) - \left[(c-h/2)^2 - 2(c-h/2)^2\right]\right\} \tag{1.22}$$

Simplify to

$$V_2 = (q/2c)(h\tan\theta + 2z_0)\{x_2^2 + 2x_2[c - (h/2)] + [c - (h/2)]^2\} \quad (1.23)$$

$$dV_2x_2 = (q/c)(h\tan\theta + 2z_0)\{x^2 + [c - (h/2)]x\}dx \quad (1.24)$$

Integrate Equation 1.24 to obtain

$$V_2x_2 = (q/6c)(h\tan\theta + 2z_0)\{2x^3 + 3x^2[c - (h/2)]\} \quad (1.25)$$

Evaluate limits and obtain

$$V_2x_2 = (q/6c)(h\tan\theta + 2z_0)\{2x_2^3 + 3x_2^2(c - h/2) - [-2(c - h/2)^3 + 3(c - h/2)^3]\} \quad (1.26)$$

Simplify to

$$V_2x_2 = (q/6c)(h\tan\theta + 2z_0)\{2x^3 + 3x^2[c - (h/2)] - [c - (h/2)]^3\} \quad (1.27)$$

Integrate for the second set of limits and obtain

$$V_2 = (2q/c)(h\tan\theta + 2z_0)\{x_2^2 + 2x_2(c - h/2)[x_2^2 - 2x_2(c - h/2)]\} \quad (1.28)$$

Simplify to

$$V_2 = (2q/c)(h\tan\theta + 2z_0)[c - (h/2)]x_2 \quad (1.29)$$

$$V_2x_2 = (q/6c)(h\tan\theta + 2z_0)\{2x_2^3 + 3x_2^2(c - h/2) - [-2x_2^3 + 3x_2^2(c - h/2)]\} \quad (1.30)$$

Simplify to

$$V_2x_2 = (q/3c)(h\tan\theta + 2z_0)2x_2^3 \quad (1.31)$$

Case 3: $(h/2 + x_2) < c < h$. One corner of the rectangular footing has negative pressure. There are two sets of limits for V_3, namely $-x_2$ to $-[c - (h/2)]$ and $-(h/2)$ to $-x_2$. The derivative for the stress volume V_3 is given by

$$dV_3 = (z_4 - z_3)\,y\,dx \quad \text{or,} \quad dV_3 = (q/c)(\cot\theta + \tan\theta)[x + (h/2)]\,\{x + [c - (h/2)]\}dx \quad (1.32)$$

Integrate the first set of limits to obtain

$$V_3 = (q/6c)(\cot\theta + \tan\theta)\{-2(c-h/2)^3 + 3c(c-h/2)^2 - 3h(c-h/2)(c-h/2)$$
$$- \left[-2x_2^3 + 3cx_2^2 - 3h(c-h/2)x_2\right]\} \tag{1.33}$$

Simplify to

$$V_3 = (q/6c)(\cot\theta + \tan\theta)\{(c-h/2)^2(2h-c) - x_2\left[2x_2^2 - 3cx_2 + 3h(c-h/2)\right]\} \tag{1.34}$$

$$V_3x_3 = (q/12c)(\cot\theta + \tan\theta)\{3(c-h/2)^4 - 4c(c-h/2)^3$$
$$+ 3h(c-h/2)^3 - 3x_2^4 - 4cx_2^3 + 3h(c-h/2)x_2^2\} \tag{1.35}$$

Simplify to

$$V_3x_3 = (q/12c)(\cot\theta + \tan\theta)\{x_2^2\left[3x_2^2 - 4cx_2 + 3h(c-h/2)\right]$$
$$-(c-h/2)^3(1.5h-c)\} \tag{1.36}$$

Integrate for the second set of limits to obtain

$$V_3 = (q/6c)(\cot\theta + \tan\theta)\{-2x_2^3 + 3cx_2^2 - 3hx_2(c-h/2)$$
$$-[-2h^3/8 + 3ch^2/4 - (3h^2/2)(c-h/2)(c-h)]\} \tag{1.37}$$

Simplify to

$$V_3 = (q/6c)(\cot\theta + \tan\theta)\{x_2\left[-2x_2^2 + 3cx_2 - 3h(c-h/2)\right] - (h^2/4)(2h-3c)\} \tag{1.38}$$

$$V_3x_3 = (q/12c)(\cot\theta + \tan\theta)\{3x_2^4 - 4cx_2^3 + 3hx_2^2(c-h/2)$$
$$- [3(h^4/16) - 4ch^3/8 + 3h^3/4)(c-h/2)]\} \tag{1.39}$$

Simplify to

$$V_3x_3 = (q/12c)(\cot\theta + \tan\theta)\{x_2^2\left[3x_2^2 - 4cx_2 + 3h(c-h/2)\right] - (h^3/16)(4c-3h)\} \tag{1.40}$$

Case 4: $c > h$. Whole area of the rectangular footing in compression. The standard flexure formula is applicable. For this case, the values of V_1, V_2, V_3, V_1x_1, V_2x_2, and V_3x_3 are given by equations 1.17, 1.29, 1.38, 1.19, 1.31, and 1.40, respectively.

Equations 1.1 to 1.40 yield the capacity of the section along the capacity axis or X-axis measured as uplift moment, M, and uplift force, P. M is the component, parallel to the capacity axis, of the resultant bending moment capacity of the section, M_R. To obtain M_R (the resultant moment capacity), M_z (the moment perpendicular to the capacity axis) should be included such that

$$M_R = \left\{ M^2 + M_z^2 \right\}^{1/2} \tag{1.41}$$

in which M is the moment component parallel to the X-axis:

$$M = (V_1x_1 + V_2x_2 + V_3x_3) \tag{1.42}$$

M_z is the moment component perpendicular to the X-axis:

$$M_z = (V_1z_1 + V_2z_2 + V_3z_3) \tag{1.43}$$

The moment around the X-axis for V_1 is given by the derivative

$$dV_1z_1 = (z_1 - z_2) \, y \, \{(1/2)(z_1 + z_2)\}dx$$

$$= \{0.50 \, V_1 \, x_1 \, (\cot - \tan) + \{z_o - 0.25h \, (\cot - \tan) \, V_1\}dx \tag{1.44}$$

Integrate to obtain

$$V_1z_1 = 0.50(\cot \theta - \tan \theta) \, V_1x_1 + \{z_o - 0.25h(\cot \theta - \tan \theta)\}V_1 \tag{1.45}$$

The moment about the X-axis for V_2 is given by the derivative

$$dV_2x_2 = (z_1 - z_3) \, y \, \{(1/2)(z_1 + z_3)\}dx = V_2x_2 \tan \theta \, x \, dx \tag{1.46}$$

Integrate to obtain

$$V_2z_2 = V_2x_2 \tan \theta \tag{1.47}$$

The moment around the X-axis for V_3 is given by the derivative

$$dV_3z_3 = (z_4 - z_3)y\{(1/2)(z_4 + z_3)\}dx$$

$$= [- 0.50 \, (\cot - \tan) \, V_3x_3 + \{0.25 \, h \, (\cot - \tan) - z_o\}]dx \tag{1.48}$$

Integrate to obtain

$$V_3z_3 = -0.50(\cot \theta - \tan \theta) \, V_3x_3 + \{0.25h(\cot \theta - \tan \theta) - z_o\} \tag{1.49}$$

$$V = (V_1 + V_2 + V_3) \tag{1.50}$$

The above equations are sufficient to determine the capacity curve when θ has a value from zero to α for normal design conditions, that is, when the resultant values of external vertical loads fall within the sector defined by zero and α. However, when θ has a value from α to $\pi/2$ for conditions that are unusual or accidental or to check values for theoretical considerations, the following equations should be utilized.

When $\alpha < \theta < (\pi/2 - \alpha)$,

$$z_1 = -\tan \theta \, (x - h/2) - z_o \tag{1.51}$$

$$z_2 = \cot \theta \, (x - h/2) - z_o \tag{1.52}$$

$$z_3 = -\tan \theta \, (x + h/2) + z_o \tag{1.53}$$

$$z_4 = \cot \theta \, (x + h/2) + z_o \tag{1.54}$$

As a result of the above changes, multiply values for V_2 and V_2x_2 by $(h \tan \theta - 2z_o)$ instead of $(h \tan \theta + 2z_o)$ in the previous derivations to obtain for the first set of limits:

$$V_2 = (q/2c)(h \tan \theta - 2z_o)\left\{x_2^2 + 2x_2[c - (h/2)] + [c - (h/2)]^2\right\} \tag{1.55}$$

$$V_2x_2 = (q/6c)(h \tan \theta - 2z_o)\{2x^3 + 3x^2[c - (h/2)] - [c - (h/2)]^3\} \tag{1.56}$$

and the second set of limits:

$$V_2 = (2q/c)(h \tan \theta - 2z_o)[c - (h/2)]x_2 \tag{1.57}$$

$$V_2x_2 = (q/3c)(h \tan \theta - 2z_o)2x_2^3 \tag{1.58}$$

When $(\pi/2 - \alpha) < \theta < \pi/2$,

$$dV_2 = (z_4 - z_2) \, ydx = (h \cot + 2z_o)(q/c)\{x + [c - (h/2)]\}dx \tag{1.59}$$

Integrating for the first set of limits,

$$V_2 = (q/c)(h \cot \theta + 2z_o)(1/2)\left\{x_2^2 + 2x_2(c - h/2) - [(c - h/2)^2 - 2(c - h/2)^2]\right\} \tag{1.60}$$

$$V_2x_2 = (q/6c)(h \cot \theta + 2z_o)\{2x^3 + 3x^2[c - (h/2)] - [c - (h/2)]^3\} \tag{1.61}$$

Integrate for the second set of limits; use the previous results but change the multiplier $(h \tan \theta + 2z_0)$ to $(h \cot \theta + 2z_0)$ to obtain

$$V_2 = (2q/c)(h \cot \theta + 2z_0)[c - (h/2)]x_2 \qquad (1.62)$$

$$V_2 x_2 = (q/3c)(h \cot \theta + 2z_0)2x_2^3 \qquad (1.63)$$

1.2.2 Circular footing

Figure 1.2 shows a circular footing with part of its area in tension. Label the variable parameters R and q. The equation of the circle is given from analytic geometry as

$$z^2 + x^2 = R^2 \qquad (1.64)$$

The equation of the stress diagram is given by the expression

$$y = (q/c)\{x + (c - R)\} \qquad (1.65)$$

The vertical uplift force is designated as V and the moment uplift as M. The compressive depth is represented by c. The external load is represented in Figure 1.2 as V_u located at a distance e from the centerline.

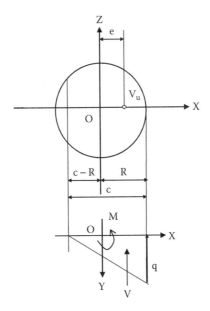

Figure 1.2 Circular footing.
Source: Jarquio, Ramon V. *Analytical Method in Reinforced Concrete.* Universal Publishers, Boca Raton, FL. p. 139. Reprinted with Permission.

There are three cases for the position of the compressive depth c in the development of the footing capacity curve for a circular section. These cases are as follows:

Case 1: $0 < c < R$. The derivative of the stress V is given by the expression

$$dV = (2q/c)\{x + c - R\}(R^2 - x^2)^{1/2} \, dx \tag{1.66}$$

Integrate using the following standard formulas and evaluate from the limits $(R - c)$ to R.

$$(R^2 - x^2)^{1/2} \, dx = (1/2)\{x(R^2 - x^2)^{1/2} + R^2 \arcsin(x/R)\} \tag{1.67}$$

$$(R^2 - x^2)^{1/2} \, x \, dx = -(1/3)\{(R^2 - x^2)^3\}^{1/2} \tag{1.68}$$

$$(R^2 - x^2)^{1/2} \, x^2 \, dx = -(x/4)\{(R^2 - x^2)^3\}^{1/2} +$$
$$(R^2/8) \{x \, R^2 - x^2)^{1/2} + R^2 \arcsin(x/R)\} \tag{1.69}$$

Hence,

$$V = (2q/c)(1/3)\{(2Rc - c^2)^3\}^{1/2} - (2q/c)(1/2)(R - c) \{1.5708R^2 -$$
$$(R - c)(2Rc - c^2)^{1/2} - R^2 \arcsin[(R - c)/R]\} \tag{1.70}$$

The expression for the derivative of the bending moment around the centerline of the circular footing is

$$dM = (2q/c)\{(R^2 - x^2)^{1/2} \, x^2 \, dx + (c - R) \, (R^2 - x^2)^{1/2} \, x \, dx\} \tag{1.71}$$

Integrate and evaluate from limits $(R - c)$ to R and obtain

$$M = (2q/c)(1/3)\{(2Rc - c^2)^3\}^{1/2}\}(c - R) + (2q/c)\{(\pi R^4/16 +$$
$$(1/4) \, (R - c)[(2Rc - c^2)^3]^{1/2} - (1/8)R^2[(R - c)(2Rc - c^2)^{1/2} +$$
$$R^2 \arcsin(R - c)/R]\} \tag{1.72}$$

Simplify to

$$M = (2q/c)\{(\pi R^4/16) - R^2/8) \arcsin[(c - R)/R] - (R^2/8)(R - c)$$
$$\times (2Rc - c^2)^{1/2} - (1/12)(R - c) \, [(2Rc - c^2)^3]^{1/2}\} \tag{1.73}$$

Case 2: $R < c < 2R$. The derivative of the stress volume for this case is

$$dV = (2q/c)(x + c - R) \, (R^2 - x^2)^{1/2} \, dx \tag{1.74}$$

Integrate and evaluate from limits $-(c - R)$ to R to obtain

$$V = (2q/c)(1/3)\{(2Rc - c^2)^3\}^{1/2} + (2q/c)(1/2)(c - R) \{(\pi R^2/2)$$
$$+ (c - R)(2Rc - c^2)^{1/2} + R^2 \arcsin[(c - R)/R]\} \tag{1.75}$$

Simplify to get

$$V = (2q/c)\{(1/3)[(2Rc - c^2)^3]^{1/2} + (c - R)[1.5708R^2$$

$$+ (c - R)(2Rc - c^2)^{1/2} + R^2 \arcsin(c - R)/R]\} \qquad (1.76)$$

The derivative for the bending moment is given by

$$dM = (2q/c)[x^2 + (c - R)x] (R^2 - x^2)^{1/2} dx \qquad (1.77)$$

Integrate from $-(c - R)$ to R to get

$$M = (2q/c)(1/3)(c - R)\{(2Rc - c^2)^3\}^{1/2} + (2q/c)\{(\pi R^4/16) +$$

$$(R^4/8) \arcsin [(c - R)/R] - (1/4)(c - R)[(2Rc - c^2)^3]^{1/2}$$

$$+ (R^2/8)(2Rc - c^2)^{1/2} \qquad (1.78)$$

Simplify to get

$$M = (2q/c)\{(\pi R^4/16) + (R^4/8)\arcsin[(c - R)/R] +$$

$$(1/12)(c - R) \times [(2Rc - c^2)^3]^{1/2} + (R^2/8)(c - R)(2Rc - c^2)^{1/2}\} \qquad (1.79)$$

Case 3: $c > R$. The derivative for the stress volume under this condition is

$$dV = (2q/c)(x + c - R) (R^2 - x^2)^{1/2} dx \qquad (1.80)$$

Integrate and evaluate from limits $-R$ to R to obtain

$$V = (2q/c)(1/2)(c - R)(\pi R^2) \quad \text{or} \quad V = (q/c)(c - R) \pi R^2 \qquad (1.81)$$

The derivative for the bending moment is given by

$$dM = (2q/c)\{x^2 + (c - R)x\}(R^2 - x^2)^{1/2} dx \qquad (1.82)$$

Integrate and evaluate from limits $-R$ to R

$$M = (2q/c)\{(\pi R^4/16) + (\pi R^4/16)\} \quad \text{or} \quad M = (q/4c)\pi R^4 \qquad (1.83)$$

Note that because of symmetry in a circular section, there is no M_z, and hence $M_R = M$, unlike in a rectangular section.

The above equations are all we need to be able to plot the capacity curve of a circular footing with a given allowable soil-bearing pressure.

The planar distribution of stresses under the footing can also be used to analyze the capacity of any circular or rectangular steel section. We shall be using the same approach in Chapter 2 for developing the equations for these steel sections.

The reader should note that we needed only basic mathematics and our knowledge of strength of materials to solve for the capacity of a rectangular and circular section with known planar stress distribution.

Since we have represented in our solutions all the variable parameters involved in the analysis, the above derived equations can solve any rectangular or circular footing. Once our equations are set up in Excel worksheets, we can enter the dimensions of the footing and the allowable soil-bearing pressure and can plot the capacity curve.

A structural engineer has to solve two main problems. One is the determination of the internal capacity of a given section. We have solved this one by the above equations. The other problem is the determination of the external loads based on code requirements and normal practice. For this, the structural engineer uses his expertise to decide the magnitude of the external loads anticipated to act on the given rectangular or circular footing. By plotting the external loads on the capacity curve, the engineer can easily determine whether the given section is adequate to support the external loads. The engineer can label these external loads as V_u and M_u. The determination of these loads is not covered here, as this subject is governed by codes of practice such as the American Association of State Highway Transportation Officials (AASHTO), American Concrete Institute (ACI), city, state, and federal building code regulations, and other codes of practice elsewhere.

1.3   Footing capacity curve

Footing capacity curve from Equations 1.1 to 1.63 and 1.64 to 1.83 for rectangular footing and circular footing, respectively, is a plot of the vertical uplift and moment uplift forces. The vertical uplift forces, V are plotted on the vertical scale and the moment uplift forces, M, are plotted on the horizontal scale for all the ranges of values of the compressive depth c of the footing. From this plot the key points are noted, and characteristics of the key points are discussed below.

Example 1.1

A rectangular footing 10 ft. (3.05 m) by 12 ft. (3.66 m) is resting on soil with a maximum allowable bearing pressure of 7.2 kips per square foot (345 kPa). Plot the footing capacity curve when the footing capacity axis is along the diagonal.

> **Solution:** In equations 1.1 to 1.63 as applicable, substitute b = 10 ft. (3.05 m), d = 12 ft. (3.66 m), q = 7.2 kips per square foot (345 kPa), and $\theta = \alpha = \arctan(10/12)$. Figure 1.3 shows the results obtained from Microsoft Excel 97 worksheets of the above-listed equations. This is the capacity curve when the capacity axis is positioned at the diagonal. As the capacity axis is varied from zero to $\pi/2$, the capacity curve varies as well.

From Figure 1.3, the maximum resultant uplift moment, M_R is at key point S and its tabulated value is 971 kN-m. This is the maximum moment to use in determining the amount of reinforcement for the given size of footing and allowable soil bearing pressure. This valve is calculated in Equation 1.41. However, this valve should be reduced to account for the overburden on top of the footing. This net bending moment should be the basis for determining the correct amount of reinforcing bars along the width and depth of footing foundation. The net bending moment is reduced by the amount of

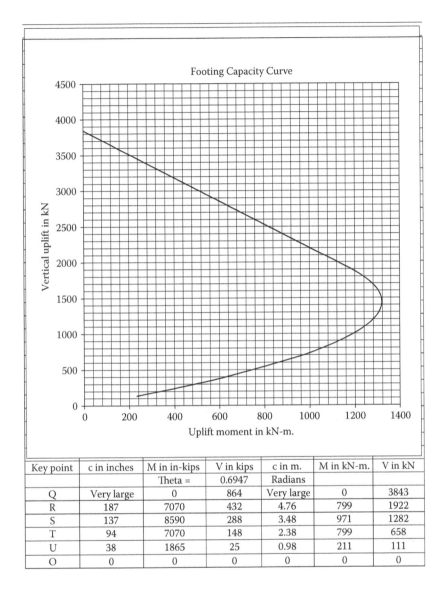

Key point	c in inches	M in in-kips	V in kips	c in m.	M in kN-m.	V in kN
		Theta =	0.6947	Radians		
Q	Very large	0	864	Very large	0	3843
R	187	7070	432	4.76	799	1922
S	137	8590	288	3.48	971	1282
T	94	7070	148	2.38	799	658
U	38	1865	25	0.98	211	111
O	0	0	0	0	0	0

Figure 1.3 Rectangular footing capacity curve.

overburden on top of the footing. This reduced bending moment will determine the amount of reinforcement along the width and depth of the footing.

The vertical uplift resistance at the point of maximum moment is approximately 1282 kN. The two most important points on the curve are the point of maximum uplift moment and the point when $c = h$, when the zero pressure is at the opposite end of the maximum pressure along the diagonal of the rectangular footing. Beyond this point, the footing area is in full compression and is commonly designed according to the standard flexure formula of adding the stress due to the vertical load and stress due to the bending moment. Note that the footing capacity curve covers all the cases when the footing area is in full compression, and in those cases external loads may be plotted to determine whether the footing is adequate to resist these external loads. When $c = h = 4.76$ m, the value of vertical uplift is 1922 kN and the uplift moment is 799 kN-m.

The footing capacity curve is for $\theta = \alpha$, which is the inclination of the diagonal with the horizontal axis 2. For other positions of θ, the footing capacity curve may be plotted as, for instance, when $\theta = 0$, which is the case for a retaining wall footing subjected to uniaxial bending moment loads.

Example 1.2

Given a circular footing 20 ft. (6.10 m) in diameter with an allowable soil-bearing pressure of 7.2 kips per square foot (345 kPa), plot the footing capacity curve and list the values at key points of the capacity curve.

Solution: In Equations 1.64 to 1.83 as applicable, substitute $R = 3.05$ m and $q = 345$ kPa. Figure 1.4 is the capacity curve obtained from Excel worksheets. We note from Figure 1.4 that when the

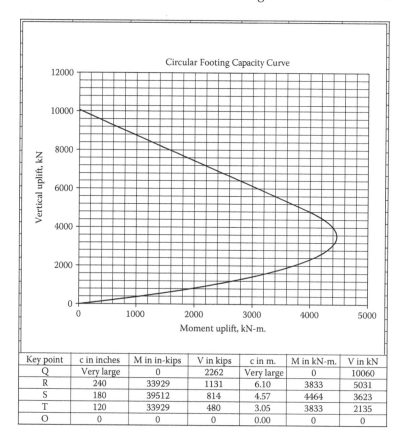

Key point	c in inches	M in in-kips	V in kips	c in m.	M in kN-m.	V in kN
Q	Very large	0	2262	Very large	0	10060
R	240	33929	1131	6.10	3833	5031
S	180	39512	814	4.57	4464	3623
T	120	33929	480	3.05	3833	2135
O	0	0	0	0.00	0	0

Figure 1.4 Circular footing capacity curve.

uplift moment is increased, the corresponding vertical uplift force is reduced. The value of the maximum uplift moment is 4464 kN-m. This is the maximum gross uplift moment on the footing. For determining the maximum steel reinforcement in the circular footing, reduce this gross moment by subtracting the overburden pressure at the top of the footing. The net bending moment will yield the maximum area of steel reinforcement to be used for this footing.

The structural design for footing depth and reinforcement will be illustrated for the rectangular footing example above, but first we will discuss the key points in the footing capacity curve.

1.3.1 Key points

Key points are designated in Figure 1.5, and the positions of these points in the footing capacity curve are plotted schematically. These key points are the envelopes of values of the footing capacity curve as the compressive depth c is varied and corresponding vertical uplift and moment uplift forces are calculated, namely:

- Key point Q is the point where the theoretical maximum vertical uplift at the center of footing is developed, and the corresponding moment uplift is zero. This has no practical application since there

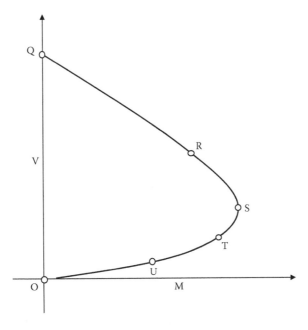

Figure 1.5 Schematic positions of key points.

is always an eccentricity of the applied vertical external loads in the design of a footing foundation.

- Key point R is the point on the capacity curve where the whole area of the rectangular footing is in compression, that is, one corner has zero uplift pressure while the other three corners have uplift pressures due to vertical loads. The envelope of values between key point R and key point Q covers the application of the standard flexure formula in the design of the footing foundation wherein the whole area of the footing is under compression.

- Key point S is the point on the capacity curve where the maximum moment uplift resistance of the footing is developed. This condition occurs when one corner of the footing is in tension. This is the gross moment value to use in the design of steel reinforcing bars for the rectangular footing foundation.

- Key point T is the point in the footing capacity curve when two corners of the rectangular footing have negative pressures; that is, part of the footing area is in tension.

- Key Point U is the point in the footing capacity curve when three corners of the rectangular footing have negative pressures.

- Key point O is the theoretical point in the capacity curve where there is no uplift pressure under the footing foundation; that is, vertical uplift and moment uplift forces are zero.

The capacity curve is shown with tabulated values at key points for the example problem. For design purposes these points are sufficient reference values with which the designer can compare the applied external vertical loads for a particular footing foundation being designed.

1.3.2   Footing design

In the design of a bridge foundation, the structural engineer determines sets of external loads consisting of resultant vertical load and bending moment. These sets of loading may be plotted on the capacity curve to determine whether these loads are inside the envelope of the capacity curve. If these loads are located inside the envelope, the particular footing is adequate for adoption in design. Otherwise, a larger footing can be tried until all the sets of loading are inside the envelope of the footing capacity curve.

The size of a footing can be directly solved when the total vertical load and biaxial bending moments are given. This method uses the standard flexure formula, which is applicable only when the whole area of the footing is in compression. However, we know from the capacity curve that the most efficient use of the footing to resist bending is when part of the footing is in tension. Under this condition it is possible to develop the maximum uplift moment to resist the applied external loads on the footing.

To design a rectangular footing with biaxial bending, we need to determine the overburden forces on top of the footing. Applying the same procedures as in our previous derivations, we can derive the equations for calculating the overburden pressure and the forces and bending moments resulting from this overburden on top of the footing.

Draw Figure 1.6 to show a circular and rectangular column on top of the footing foundation. Label the dimensional parameters of these columns. Using basic calculus, derive the equations for the forces and their corresponding bending moments.

The variables considered in the calculation of the forces due to the overburden material on top of the footing are the column dimensions and the uniform load w.

The equation of the circular column is given by

$$x^2 + z^2 = R_1^2 \tag{1.84}$$

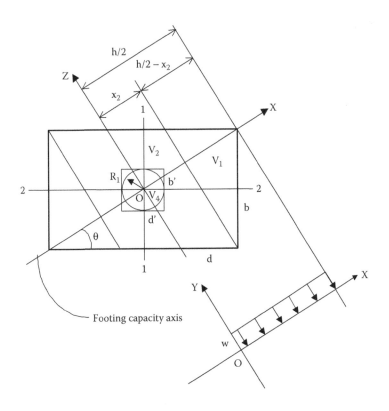

Figure 1.6 Overburden on rectangular footing.
Source: Jarquio, Ramon V. *Analytical Method in Reinforced Concrete.* Universal Publishers, Boca Raton, FL. p. 164. Reprinted with Permission.

The force due to the uniform load w is given by

$$V_4 = 2w \int \left(R_1^2 - x^2 \right)^{1/2} dx \tag{1.85}$$

$$V_4 = (1/2)w\pi R_1^2 \tag{1.86}$$

The moment of this force is given by the expression

$$V_4 x_4 = 2w \int \left\{ \left(R_1^2 - x^2 \right)^3 \right\}^{1/2} dx \tag{1.87}$$

$$V_4 x_4 = (2/3)wR_1^3 \tag{1.88}$$

The overburden forces for the rectangular footing are given by the following expressions:

$$V_1 = -w(\cot \theta + \tan \theta) \int (x - h/2)\, dx \tag{1.89}$$

$$V_1 = (1/2)(\cot \theta + \tan \theta)\{h^2/4 + x_2\,(x_2 - h)\} \tag{1.90}$$

$$V_1 x_1 = -w(\cot \theta + \tan \theta) \int \{x^2 - (h/2)x\}\, dx \tag{1.91}$$

$$V_1 x_1 = (1/12)w(\cot \theta + \tan \theta)\left\{ (h^3/4) + x_2^2(4x_2 - 3h) \right\} \tag{1.92}$$

$$V_2 = w(h \tan \theta + 2\, z_o) \int dx \tag{1.93}$$

$$V_2 = wx_2(h \tan \theta + 2z_o) \tag{1.94}$$

$$V_2 x_2 = w\,(h \tan \theta + 2\, z_o) \int x\, dx \tag{1.95}$$

$$V_2 x_2 = (1/2)w(h \tan \theta + 2z_o)x_2^2 \tag{1.96}$$

The moment due to overburden material is

$$M_o = V_1x_1 + V_2x_2 - V_4x_4 \tag{1.97}$$

When the column is rectangular, use Equations 1.89 to 1.97 by replacing d with d', b with b', h with h', x_2 with x_2', and z_o with z_o' in which b' and d' are the rectangular column dimensions. For a square column, $b' = d'$.

The moment due to overburden material is then calculated as

$$M_o = (V_1x_1 + V_2x_2) - (V_1x_1' + V_2x_2') \tag{1.98}$$

Example 1.3

Determine the total amount of reinforcing steel bars for the example rectangular footing above when the column dimension is 36 in. (0.91 m) in diameter, fc′ = 5 ksi (34.5 MPa), fy′ = 60 ksi (414 MPa), and w = 1.26 kips per square foot (60.9 kPa).

Solution:

Step 1: Determine the maximum moment uplift from the footing capacity curve for this footing. Figure 1.3 is the footing capacity curve for the allowable soil-bearing pressure of 345 kPa. From this curve we read $M_R = 971$ kN-m. This is the maximum resultant uplift moment resistance of the footing foundation when the capacity axis is along the diagonal of the rectangular footing. From the Excel worksheet, Mz = 180 kN-m and M = 954 kN-m. The inclination of the resultant moment from the horizontal axis is equal to 0.6947 - arctan(180/954) = 0.6947 − 0.1865 = 0.5082 radians (29.12°).

Step 2: Calculate the moment due to the overburden material from Equations 1.84 to 1.88 as applicable for a circular column. The moment due to overburden material for the circular column when w = 60.9 is from Equation 1.88, given by

$$V_4x_4 = (2/3)(60.9)(0.46)^3 = 3.88 \text{ kN-m}$$

Determine the moment due to the overburden material on top of rectangular footing as follows:

$$\theta = \arctan(3.05/3.66) = 39.81° \text{ or } 0.6947 \text{ radian}$$

$$h = 3.66\{\cos(39.81°)\} + 3.05\{\sin(39.81°)\} = 4.764 \text{ m}$$

$$z_o = (1/2)\{3.05[\cos(39.81°)] - 3.66[\sin(39.81°)]\} = 0$$

$$x_2 = (1/2)\{3.66[\cos(39.81°)] - 3.05[\sin(39.81°)]\} = 0.43 \text{ m}$$

$$V_2x_2 = (1/2)(60.9)\{4.764 \tan(39.81°)\}(0.43)^2 = 22.35 \text{ kN-m}$$

$$V_1x_1 = (1/12)(60.9)\{\cot(39.81° + \tan(39.81°)\}\{(1/4)(4.764)^3$$

$$+ (0.43)^2[4(0.43) - 3(4.764)]\} = 255.0 \text{ kN-m}$$

$$M_o = 255.0 + 22.35 - 3.88 = 274 \text{ kN-m}$$

Step 3: *Determine the net moment for the design of reinforcing bars. Net M = 971cos (10.68°) − 274 = 680 kN-m (6,019 in-kips) along the capacity axis. The net resultant moment is 680/cos (10.68°) = 692 kN-m. (6,125 in-kips).*

Step 4: *Solve for the total reinforcing bars and resolve the corresponding amount along the 1-1 and 2-2 axes (Jarquio, 2004, pp. 9–13).*

$$K = 49(51 + 60)(5)/(87 + 60)^2 = 1.258$$

Minimum depth to rebars

$$= \{6,125 \cos (29.12°)/(1.258)(120)\}^{1/2} = 5.95 \text{ in. (151 mm)}$$

Assume we are using 24 in. (610 mm) of total concrete thickness and 3.50 in. (90 mm) of cover to reinforcing bars. Depth to rebars = 24 − 3.50 = 20.50 in. (521 mm). c = 87(20.50)/(87 + 60) = 12.13 in. (308 mm).

$$A_{s1} = 5,351/\{60[20.50 - 0.417(12.13)]\}$$

$$= 5.78 \text{ in.}^2 \text{ (38 cm}^2) \text{ reinforcing steel bars along axis 2–2}$$

$$A_{s2} = 2,981/\{60[20.50 - 0.417(12.13)]\}$$

$$= 3.22 \text{ in.}^2 \text{ (21 cm}^2) \text{ reinforcing steel bars along axis 1–1.}$$

The total equivalent area of reinforcement = $\{(5.78)^2 + (3.22)^2\}^{1/2} = 6.62$ in.^{2} (43 cm^2), which is equal to $6,125/\{60[20.50 - 0.417(12.13)] = 6.62$ in.^{2} (43 cm^2). This amount of steel reinforcement is the maximum to put into the rectangular footing. Any more than this is a waste for the given allowable soil-bearing pressure and size of the rectangular footing in the above example.

The punching shear force can be calculated at key point R in which the value of V = 1922 kN less the overburden of 60.9{(3.05)(3.66) − 3.1416(0.46)²} to obtain 1922 − 639 = 1283 kN.

Example 1.4

Determine the amount of reinforcing steel in the circular footing example above if the column is 48 in. (1.22 m) in diameter with the same amount of overburden and stresses as the rectangular footing example.

Solution:

Step 1: Determine the maximum uplift moment from the capacity curve as 4464 kN-m. Since a circular section is symmetrical, any diameter can be a capacity axis. Hence, $M_R = M$.

Step 2: Calculate the overburden moment as follows:

$$V_4 x_4 = (2/3)(60.9)(0.61)^3 = 9.20 \text{ kN-m. for the column}$$

$$V_4 x_4 = (2/3)(60.9)(3.05)^3 = 1{,}152 \text{ kN-m}$$

$$\text{Net overburden} = 1{,}152 - 9.20 = 1{,}143 \text{ kN-m}$$

Step 3: Determine the net moment to design the reinforcing steel bars.

Net design moment $= 4{,}464 - 1{,}143 = 3{,}321$ kN-m. (29,395 in-Kips)

Step 4: Solve for the amount of reinforcing steel bars. Use a steel mesh formation with equal amounts in either direction.

Minimum depth to rebars

$$= \{29{,}395/(1.258(192)\}^{1/2} = 11.04 \text{ in. (280 mm)}$$

Assume we are using 30 in. depth of footing and 4 in. of cover to the center of the rebars. The depth to rebars is 30 − 4 = 26 in.

$$c = 87(26)/(87 + 60) = 15.388 \text{ in. (391 mm)}$$

$$A_s = 29{,}351/\{60[26 - 0.417(15.388)]\} = 24.98 \text{ in.}^2 \text{ (161 cm}^2\text{)}$$

This is the maximum amount of steel reinforcing bars in one direction and the same amount in the perpendicular direction. At any direction θ, the amount of reinforcement provided is equal to

$$A_s \cos\theta + A_s \sin\theta = A_s (\cos\theta + \sin\theta) \tag{1.99}$$

Equation 1.99 is always greater than unity, and hence adequate reinforcement is provided throughout the area of the circular footing foundation. Table 1.1 shows the numerical proof of Equation 1.99

The punching shear force can be calculated as 5031 − {0.50(60.9) (3.1416)[(3.05)² − (0.61)²]} = 4177 kN. Figure 1.7 shows the schematics of the layout of the reinforcing steel bars for the example footings.

Table 1.1 Proof of Equation (99) => 1.00

theta	cos theta	sin theta	SUM
0.001	1.0000	0.0010	1.0010
0.100	0.9950	0.0998	1.0948
0.200	0.9801	0.1987	1.1787
0.300	0.9553	0.2955	1.2509
0.400	0.9211	0.3894	1.3105
0.500	0.8776	0.4794	1.3570
0.600	0.8253	0.5646	1.3900
0.700	0.7648	0.6442	1.4091
0.800	0.6967	0.7174	1.4141
0.900	0.6216	0.7833	1.4049
1.000	0.5403	0.8415	1.3818
1.100	0.4536	0.8912	1.3448
1.200	0.3624	0.9320	1.2944
1.300	0.2675	0.9636	1.2311
1.400	0.1700	0.9854	1.1554
1.500	0.0707	0.9975	1.0682
1.571	0.0001	1.0000	1.0001

1.3.3   Centroid of a pair of V and M_R capacity values

From Figure 1.8 we can see that the location of the resultant load is determined with reference to the capacity axis. The coordinates of this location with respect to the capacity axis are as follows:

$$e_1 = M/V \quad \text{and} \quad e_2 = M_z/V \tag{1.100}$$

$$e = \left(e_1^2 + e_2^2\right)^{1/2} \tag{1.101}$$

At the key point S, where the maximum uplift moment is developed, we can test the above equations to locate the centroid of this pair of values. From the data used in our example rectangular footing we have the following:

$$e_1 = 954/1282 = 0.74 \text{ m} \quad \text{and} \quad e_2 = 180/1282 = 0.14\text{m}$$

$$e = \{(0.74)^2 + (0.14)^2\}^{1/2} = 0.757\text{m}$$

$$\beta = \arctan(180/954) = 10.68°$$

$$\theta_u = \alpha - \beta \tag{1.102}$$

When the external load is plotted inside the envelope of the capacity curve, we can be certain that the given allowable soil-bearing stress will not be exceeded. We use the proportionality of the stress diagram to calculate the actual stress at the corner of the footing as

$$q_a = (M_u/M_R)q \tag{1.103}$$

For this case we are certain that $M_R > M_u$, so the footing design is adequate for our purpose. This is the only way we can show that equilibrium

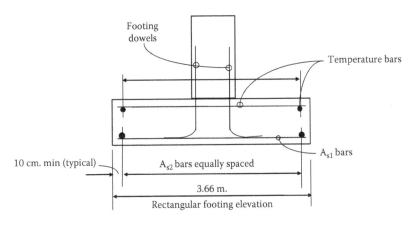

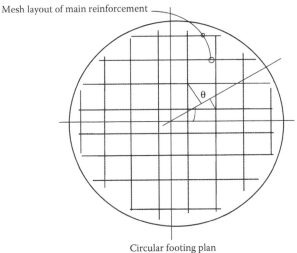

Circular footing plan

Figure 1.7 Schematic layout of reinforcing steel bars.

is satisfied when the location of M_u falls directly on the curve line. If it is inside, we apply Equation 1.103. If it is outside the envelope, we should try a larger footing until we are satisfied that the external load is inside the capacity curve. The current practice does not have this capability for demonstration and is thus unreliable.

1.3.4 Variation of the footing capacity curve

For a circular section, there will be no variation of the footing capacity because of the symmetry of a circular curve. However, for a rectangular footing section, changing the position of the capacity axis from zero to $\pi/2$ will generate different capacity curves.

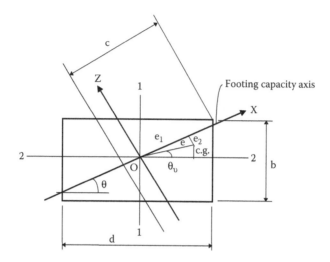

Figure 1.8 Centroid of the resultant load.
Source: Jarquio, Ramon V. *Analytical Method in Reinforced Concrete.* Universal Publishers, Boca Raton, FL. p. 132. Reprinted with Permission.

To see this variation, we will analyze values of the capacity curve for different values of the inclination of the capacity axis. Using our previous example and our Excel worksheets, Figure 1.9 and Figure 1.10 are the plots of the capacity curve at $\theta = 0$ and $\theta = \pi/2$. The variation of moment uplift at key point S as, is varied from zero to $\pi/2$, can be constructed from the data in the Excel worksheets by using the data in Figure 1.11 to test the accuracy of the standard interaction formula for biaxial bending when applied to a footing foundation.

Recall that the flexure formula is the basis of the standard interaction formula that simply assumes that the sum of the ratios of the bending moments around orthogonal axes should not exceed one or unity. These ratios relate the external moments to the internal capacity of the footing foundation. We can check the deficiency of the standard interaction formula for biaxial bending at key point S as shown in Table 1.2.

To construct the tabulation, we assume that an external load equal to M_R is acting on the rectangular footing. M_1 is the uplift moment capacity around axis 1–1, and M_2 is the uplift moment capacity around axis 2–2. The inclination of the resultant around the horizontal axis is $\theta u = \alpha - \beta$, in which $\alpha = \theta$ above and $\beta = \arctan (M_z/M)$. $M_1 = M_R \cos \theta u$ and $M_2 = M_R \sin \theta u$. From these relationships, we can now compute R_1, R_2, and R_3 to test the accuracy of the standard interaction formula for biaxial bending.

The standard interaction formula for biaxial bending and thrust is given by the expression

$$(F_a/F_a') + (F_{b1}/F_{b1}') + (F_{b2}/F_{b2}') \le 1 \qquad (1.104)$$

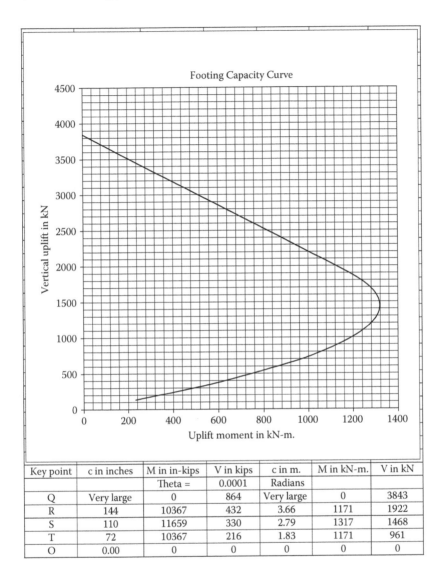

Figure 1.9 Capacity curve when theta = 0.

The terms in the numerator represent the calculated external loads due to dead loads, live loads, wind, impact, earthquake, water, and so on and multipliers that may include safety factors as per codes or at the discretion of the structural engineer.

The terms in the denominator represent the capacity of the section that can be based on allowable or ultimate stress. This interaction formula is based on the familiar flexure formula written as

$$F = (P/A) + (Mc/I) \tag{1.105}$$

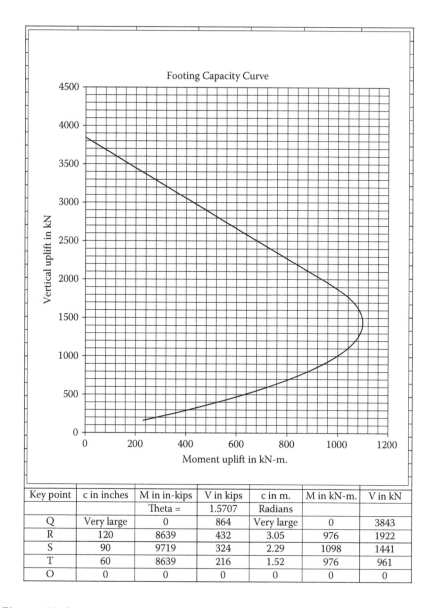

Key point	c in inches	M in in-kips	V in kips	c in m.	M in kN-m.	V in kN
		Theta =	1.5707	Radians		
Q	Very large	0	864	Very large	0	3843
R	120	8639	432	3.05	976	1922
S	90	9719	324	2.29	1098	1441
T	60	8639	216	1.52	976	961
O	0	0	0	0	0	0

Figure 1.10 Capacity curve when theta = $\pi/2$.

This is only valid when the stress in the section is compressive. When any part of the section is in tension, the flexure formula and the standard interaction formula cannot be used. Yet it is being used in analyzing biaxial bending and thrust in many codes of practice such as the ACI, ASCE, American Institute of Steel Construction (AISC), and AASHTO.

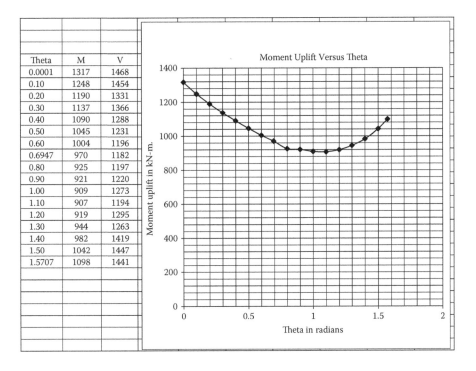

Theta	M	V
0.0001	1317	1468
0.10	1248	1454
0.20	1190	1331
0.30	1137	1366
0.40	1090	1288
0.50	1045	1231
0.60	1004	1196
0.6947	970	1182
0.80	925	1197
0.90	921	1220
1.00	909	1273
1.10	907	1194
1.20	919	1295
1.30	944	1263
1.40	982	1419
1.50	1042	1447
1.5707	1098	1441

Figure 1.11 Uplift moment variation versus theta.

Table 1.2 indicates that if we use the standard interaction formula and substitute the above calculated ratios for biaxial bending for a footing foundation, we can only capture 64% of the potential capacity of the example rectangular footing with a given allowable soil-bearing pressure. This kind of comparison of the analytical method with the current methodology will also be shown in succeeding chapters of this book to highlight the mediocrity of the standard interaction formula for biaxial bending in structural analysis.

Figure 1.12 shows the trace line of key point S, which is the point of maximum uplift moment on the footing. This shows that the shape of the bulb of the surface generated by the capacity curves from zero to $\pi/2$, or the first quadrant undulates, and the bulge (at key point S) rises and dips. The position of the bulge at zero is clearly at a higher level than at $\pi/2$, as can be seen from the chart and plot of the trace line.

The reader should note that the examples are provided to show only how the equations are applied to solving structural problems in the design of footing foundations. The examples do not necessarily represent actually constructed footing foundations.

Table 1.2 Limitations of the Standard Interaction Formula at Key Point S of Capacity Curve

M =	954	Mz =	180	M1 =	1318	Beta =	0.1865		
MR =	971	Theta =	0.6947	M2 =	1098	Theta U =	0.5082		
V =	1182	Vmax =	3843	R3 = V/Vmax	0.308	R1 + R2 + R3 =< 1.00			
M1'	M2'	MR'	R1 = M1'/M1	R2 = M2'/M2	R1 + R2	R3	Interaction	R	%
848	473	971	0.644	0.430	1.074	0.308	1.382	1.000	100
786	438	900	0.597	0.399	0.995	0.308	1.303	0.927	93
764	426	875	0.580	0.388	0.968	0.308	1.275	0.901	90
743	414	850	0.563	0.377	0.940	0.308	1.248	0.875	88
721	401	825	0.547	0.366	0.912	0.308	1.220	0.850	85
699	389	800	0.530	0.355	0.885	0.308	1.192	0.824	82
677	377	775	0.514	0.343	0.857	0.308	1.165	0.798	80
655	365	750	0.497	0.332	0.830	0.308	1.137	0.772	77
633	353	725	0.481	0.321	0.802	0.308	1.109	0.747	75
612	341	700	0.464	0.310	0.774	0.308	1.082	0.721	72
590	328	675	0.447	0.299	0.747	0.308	1.054	0.695	70
568	316	650	0.431	0.288	0.719	0.308	1.026	0.669	67
546	**304**	**625**	**0.414**	**0.277**	**0.691**	**0.308**	**0.999**	**0.644**	**64**
524	292	600	0.398	0.266	0.664	0.308	0.971	0.618	62
502	280	575	0.381	0.255	0.636	0.308	0.944	0.592	59
480	268	550	0.365	0.244	0.608	0.308	0.916	0.566	57
459	255	525	0.348	0.233	0.581	0.308	0.888	0.541	54
437	243	500	0.331	0.222	0.553	0.308	0.861	0.515	51
415	231	475	0.315	0.211	0.525	0.308	0.833	0.489	49
393	219	450	0.298	0.199	0.498	0.308	0.805	0.463	46
371	207	425	0.282	0.188	0.470	0.308	0.778	0.438	44
349	195	400	0.265	0.177	0.442	0.308	0.750	0.412	41
328	182	375	0.249	0.166	0.415	0.308	0.722	0.386	39
306	170	350	0.232	0.155	0.387	0.308	0.695	0.360	36
284	158	325	0.215	0.144	0.359	0.308	0.667	0.335	33
262	146	300	0.199	0.133	0.332	0.308	0.639	0.309	31
240	134	275	0.182	0.122	0.304	0.308	0.612	0.283	28

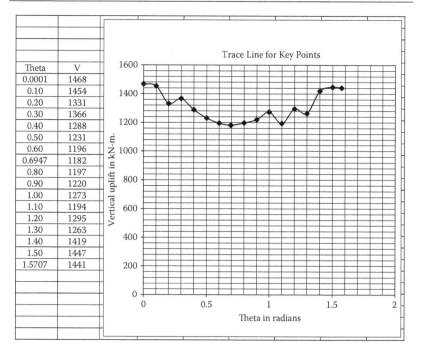

Theta	V
0.0001	1468
0.10	1454
0.20	1331
0.30	1366
0.40	1288
0.50	1231
0.60	1196
0.6947	1182
0.80	1197
0.90	1220
1.00	1273
1.10	1194
1.20	1295
1.30	1263
1.40	1419
1.50	1447
1.5707	1441

Trace Line for Key Points

Vertical uplift in kN-m.

Theta in radians

Figure 1.12 Trace of the bulge at key point S of the capacity curves.

1.4 Surface loading

For more than half a century, the integration of Boussinesq's elastic equation for a rectangular and trapezoidal surface loading has eluded investigators. W. Steinbrenner (1936) first presented the integration of the elastic equation for rectangular or uniform loading. His solution, however, was only halfway done and hence, no better than the Newmark charts since mechanical integration has to be performed to get the total vertical stress. His formulas for the vertical stress at any depth appeared in some textbooks without his name mentioned. He indicated in his paper that the integration for total vertical stress became difficult. Terzaghi (1943) probably read Steinbrenner's paper and concluded in his book that the equation cannot be simplified. Hence, graphical and finite element procedures have been the standard tools for calculating vertical stresses in soil mechanics.

The following derivations will show the results of the integration of the elastic equation for vertical stress using standard integration formulas. Three levels of substitutions have been employed to solve these problems. The reader should be able to follow the integration of the elastic equation, although some intermediate steps in the derivation are not shown.

1.4.1 Uniform load on a rectangular area

The vertical stress under a point load through a soil medium is given by Boussinesq's equation as

$$\sigma_z = (3Q/2\pi)\{z^3/(z^2 + r^2)^{5/2}\} \qquad (1.106)$$

For a uniformly loaded rectangular area of dimensions a and b, performing the required integration of Equation 1.104 can derive the expression for the vertical stress h deep under a corner. Note that Terzaghi (1943) concluded that the integration could not be done.

In Figure 1.13, let $\sigma_z = P$; $z = h$; $r^2 = x^2 + y^2$; and $dQ = q\,dxdy$. Substitute these expressions in Equation 1.106 and write

$$dP = (3dQ/2\pi)\{h^3/(h^2 + x^2 + y^2)^{5/2}\} \qquad (1.107)$$

Integrate Equation 1.107 and write

$$P = (3qh^3/2\pi)\int_0^b \int_0^a \{dxdy/(h^2 + x^2 + y^2)^{5/2}\} \qquad (1.108)$$

Let $m^2 = h^2 + y^2$, $x = m\tan z$, $x^2 = m^2\tan^2 z$, and $dx = m\sec^2 z\,dz$. Substitute these expressions in Equation 1.108 and integrate with respect to x. The equation obtained is the vertical stress at a point h deep below the end of a finite line load, that is,

$$P = (3qh^3/2\pi)\int_0^a \{dx/(h^2 + y^2 + x^2)^{5/2}\} \qquad (1.109)$$

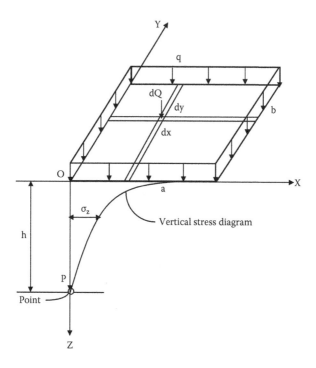

Figure 1.13 Vertical stress under a uniform load intensity.
Source: Jarquio, Ramon V. *Analytical Method in Reinforced Concrete.* Universal Publishers, Boca Raton, FL. p. 171. Reprinted with Permission.

$$P = (3qh^3/2\pi) \int (m\sec^2 z/m^5 \sec^5 z)dz \qquad (1.110)$$

Equation 1.110 when integrated is reduced to

$$P = (3qh^3/2\pi m^4) \{\sin z - (\sin^3 z/3)\} \qquad (1.111)$$

Equation 1.111 is reverted back as a function of x as follows:

$$P = (3qh^3/2\pi)[1/(h^2 + y^2)]\{[x/(h^2 + y^2 + x^2)^{1/2}] - [x^3/3(h^2 + y^2 + x^2)^{3/2}]\} \qquad (1.112)$$

Evaluate Equation 1.112 using the limits from 0 to a and simplify to

$$P = (qah^3/2\pi)\{[(2a^2 + 3h^2) + 3y^2]/(h^2 + y^2)^2(h^2 + a^2 + y^2)^{3/2}\} \qquad (1.113)$$

To find the vertical stress at the corner of the rectangular area, integrate Equation 1.113 with respect to y such that

$$P = (qah^3/2\pi) \int_0^b \{[(2a^2 + 3h^2) + 3y^2]/(h^2 + y^2)^2(h^2 + a^2 + y^2)^{3/2}\}dy \qquad (1.114)$$

Let

$$h^2 + a^2 + y^2 = z^2 y^2; \quad z = (1/y)(h^2 + a^2 + y^2)^{1/2}; \quad h^2 + a^2 = y^2(z^2 - 1)$$

$$h^2 + a^2 + y^2 = [z^2/(z^2 - 1)](h^2 + a^2)$$

$$y^2 = (h^2 + a^2)/(z^2 - 1); \quad (h^2 + a^2 + y^2)^{1/2} = (h^2 + a^2)^{1/2}[z/(z^2 - 1)^{1/2}];$$

$$y = (h^2 + a^2)^{1/2}/(z^2 - 1)^{1/2}$$

$$dy = - \{(h^2 + a^2)^{1/2}/(z^2 - 1)^{3/2}\}z\,dz$$

Substitute these expressions in Equation 1.114 and integrate first term as

$$I = - (qah^3/2\pi)(2a^2 + 3h^2)\int \{(z^2 - 1)^2/h^4z^2[z^2 + (a^2/h^2)]^2\}dz \qquad (1.115)$$

Expand Equation 1.115 to

$$I = (qah^3/2\pi)(2a^2 + 3h^2)\left\{-\int z^2 dz/[z^2 + (a^2/h^2)]^2 \right.$$

$$\left. + 2\int dz/[z^2 + (a^2/h^2)]^2 - \int dz/z^2[z^2 + (a^2/h^2)]^2 \right\} \qquad (1.116)$$

Let $z = (a/h) \tan \alpha$, $dz = (a/h) \sec^2 \alpha \, d\alpha$, and $z^2 = (a^2/h^2) \tan^2 \alpha$. Substitute these expressions in Equation 1.116 and integrate terms separately as follows:

$$-\int z^2 \, dz/[z^2 + (a^2/h^2)]^2 = - (h/a)\int (\tan^2\alpha/\sec^2\alpha)d\alpha \qquad (1.117)$$

Equation 1.117 is integrated as

$$-\int z^2 \, dz/[z^2 + (a^2/h^2)]^2 = - (h/2a)(\alpha - \sin \alpha \cos \alpha) \qquad (1.118)$$

$$2\int dz/[z^2 + (a^2/h^2)]^2 = (2h^3/a^3)\int \cos^2 \alpha \, d\alpha \qquad (1.119)$$

Equation 1.119 is integrated to

$$2\int dz/[z^2 + (a^2/h^2)]^2 = (h^3/a^3)(\alpha + \sin \alpha \cos \alpha) \qquad (1.120)$$

$$-\int dz/[z^2 + (a^2/h^2)]^2 = - (h^5/a^5)\int \{\cos^4 \alpha/\sin^2 \alpha\} \, d\alpha \qquad (1.121)$$

Integrate Equation 1.121 to obtain

$$-\int dz/[z^2 + (a^2/h^2)]^2 = (h^5/2a^5)\{2\cot\alpha + 3\alpha + \sin\alpha\cos\alpha\} \quad (1.122)$$

$$\text{II} = 3(qah^3/2\pi)\int\{y^2/(h^2 + y^2)^2(h^2 + a^2 + y^2)^{3/2}\}dy \quad (1.123)$$

Equation 1.123 can be transformed as a function of z such that

$$3(qah^3/2\pi)\int\{y^2/(h^2 + y^2)^2(h^2 + a^2 + y^2)^{3/2}\}dy = (3qah^3/2\pi h^4)$$

$$\left\{-\int dz/[z^2 + (a^2/h^2)]^2 + \int dz/z^2[z^2 + (a^2/h^2)]^2\right\} \quad (1.124)$$

Integrate the terms in Equation 1.124 separately:

$$-\int dz/[z^2 + (a^2/h^2)]^2 = -(h^3/a^3)\int \cos^2\alpha\, d\alpha \quad (1.125)$$

Integrate Equation 1.125 to

$$-\int dz/[z^2 + (a^2/h^2)]^2 = -(h^3/2a^3)\{\alpha + \sin\alpha + \sin\alpha\cos\alpha\} \quad (1.126)$$

$$\int dz/z^2[z^2 + (a^2/h^2)]^2 = (h^5/a^5)\int\{\cos^4\alpha/\sin^2\alpha\}d\alpha \quad (1.127)$$

Equation 1.127 is integrated to

$$\int dz/z^2[z^2 + (a^2/h^2)]^2 = -(h^5/2a^5)\{2\cot\alpha + 3\alpha + \sin\alpha\cos\alpha\} \quad (1.128)$$

For set I let

$$K_1 = (3qah^3/2\pi h^4)\{(2a^2 + 3h^2)/(h^2 + a^2)\} \quad (1.129)$$

For set II let

$$K_2 = (3qah^3/2\pi h^4) \quad (1.130)$$

Collect terms involving α, $\sin\alpha\cos\alpha$, and $\cot\alpha$ as follows:

- α terms:

$$K_1\{[(h^3/a^3) + (3h^5/2a^5)] - (h/2a)\}$$
$$- K_2\,[(h^3/2a^3) + (3h^5/2a^5)] + (1/2a^5)\,\{K_1(2a^2h^3 + 3h^5 - a^4h)$$
$$- K_2\,(a^2h^3 + 3h^5) + (qa/4\pi a^5h)\,\{[(2a^2 + 3h^2)(2a^2h^3 + 3h^5 - a^4h)/(h^2 + a^2)]$$
$$- 3\,(a^2h^3 + 3h^5)\} = -\,(q\alpha/2\pi) \tag{1.131}$$

- $\sin\alpha\cos\alpha$ terms:

$$K_1\,[(h/2a) + (h^3/a^3) + (h^5/2a^5)] - K_2\,[(h^3/2a^3) + (h^5/2a^5)] +$$
$$(q/4\pi a^4h)\,[1/(h^2 + a^2)]\{(2a^2 + 3h^2)(a^4h + 2a^2h^3 + h^5) -$$
$$3(h^2 + a^2)(a^2h^3 + h^5)\} = (q/2\pi a^2)(h^2 + a^2)\sin\alpha\cos\alpha \tag{1.132}$$

- $\cot\alpha$ terms:

$$(qh^4\cot\alpha/2\pi a^4)[1/(h^2 + a^2)]\,\{2a^2 + 3h^2 -$$
$$(3h^2 + 3a^2)\} = -\,qh^4\cot\alpha/2\pi a^2\,(h^2 + a^2) \tag{1.133}$$

Transform Equations 1.131, 1.132, and 1.133 to the variable y. Use the relationships $\cot\alpha = a/hz$, $\alpha = \arctan(hz/a)$, and $\sin\alpha\cos\alpha = ahz/(h^2z^2 + a^2)$.

$$\text{Sum } (1.131 + 1.132 + 1.133) = -\,(q/2\pi)\arctan(hz/a)$$
$$+ (q/2\pi a^2)(h^2 + a^2)\,[ahz/(h^2z^2 + a^2)]$$
$$- qah^4/2\pi a^2hz(h^2 + a^2) \tag{1.134}$$

Hence,

$$P = (q/2\pi)\{-\arctan(hz/a) + hz(h^2 + a^2)/a(h^2z^2 + a^2)\,h^3/az(h^2 + a^2)\} \tag{1.135}$$

Substitute $z = \{(1/y)(h^2 + a^2 + y^2)^{1/2}\}$ in Equation 1.135 to obtain

$$P = (q/2\pi)\{-\arctan[h(h^2 + a^2 + y^2)^{1/2}/ay]$$
$$+ hy(h^2 + a^2)(h^2 + a^2 + y^2)^{1/2}/a\,[h^2\,(h^2 + a^2 + y^2) + a^2y^2]\} \tag{1.136}$$

Evaluate Equation 1.136 when the limit of y is from 0 to b to yield

$$P = (q/2)\{bh(h^2 + a^2)(h^2 + a^2 + b^2)/a[h^2(h^2 + a^2 + b^2) + a^2b^2] - bh^3/[a(h^2 + a^2)$$
$$\times\,(h^2 + a^2 + b^2)^{1/2} - \arctan[(h/ab)(h^2 + a^2 + b^2)^{1/2}] - (-/2)\} \tag{1.137}$$

Simplify Equation 1.137 to

$$P = (q/2\pi)\{abh[(h^2 + a^2) + (h^2 + b^2)]/(h^2 + a^2)(h^2 + b^2)(h^2 + a^2 + b^2)^{1/2}$$

$$+ (\pi/2) - \arctan[(h/ab)(h^2 + a^2 + b^2)^{1/2}]\} \qquad (1.138)$$

Equation 1.138 is the formula for the vertical stress h deep under the corner of a rectangular area loaded with a uniform load q.

W. Steinbrenner has integrated Boussinesq's elastic equation of a point load up to Equation 1.138. However, he could not proceed any further and concluded that the integration for the total vertical stress became very difficult. His equation expressed the last two terms of Equation 1.138 as

$$\arctan[ab/h(h^2 + a^2 + b^2)^{1/2}]$$

His equation appeared in the book by H. G. Poulous and E. H. Davis entitled *Elastic Solutions for Soil and Rock Mechanics* (1974).

Steinbrenner's equation is no better than Newmark's chart since Equation 1.138 has to be applied repeatedly at intervals throughout depth h to get the total vertical stress by mechanical integration such as the trapezoidal rule of summing areas.

Terzaghi (1943) concluded that the integration of Boussinesq's elastic equation cannot be done, we show in the following derivation that his conclusion was premature.

We can also write Equation 1.138 as

$$P = (q/2\pi)\{[abh(A^2 + B^2)/A^2B^2C] + (\pi/2) - \arctan(hC/ab)\} \qquad (1.139)$$

in which $A = (h^2 + a^2)^{1/2}$; $B = (h^2 + b^2)^{1/2}$; $C = (h^2 + a^2 + b^2)^{1/2}$.

It can be seen from Equation 1.139 that the rectangular area can be rotated around the corner in any position without changing the value of the vertical stress; that is, a and b are interchangeable. Moreover, when h is made equal to z, Equation 1.139 becomes the equation of the vertical stress along the Z-axis. From this, the integration for the total vertical stress can be done as shown in the following derivations. Let

$$M = ab \int zdz/(z^2 + b^2)(z^2 + a^2 + b^2)^{1/2} \qquad (1.140)$$

and

$$u = (z^2 + a^2 + b^2)^{1/2}; \quad u^2 = z^2 + a^2 + b^2; \quad z = [u^2 - (a^2 + b^2)]^{1/2};$$

$$dz = udu/[u^2 - (a + b^2)]^{1/2}$$

Substitute these expressions in Equation 1.140 to obtain

$$M = ab \int du/(u^2 - a^2) \qquad (1.141)$$

Integrate Equation 1.141 to yield

$$M = (b/2)Ln\{(u - ba)/(u + a)\} \tag{1.142}$$

Let

$$N = ab \int zdz/(z^2 + a^2)(z^2 + a^2 + b^2)^{1/2} \tag{1.143}$$

Similarly, Equation 1.143 is integrated to

$$N = (a/2) Ln\{(u - b)/(u + b)\} \tag{1.144}$$

Let

$$O = \int \arctan\{(z/ab)(z^2 + a^2 + b^2)^{1/2}\}dz \tag{1.145}$$

Integrate Equation 1.143 by parts. Let

$$u = \arctan \{(z/ab)(z^2 + a^2 + b^2)^{1/2}\}; \ v = \int dz$$

$$uv = z \arctan\left\{(z/ab)(z^2 + a^2 + b^2)^{1/2}\right.$$

$$\left. - \int zab(z^2 + z^2 + a^2 + b^2)dz/ [a^2b^2 + z^2(z^2 + a^2 + b^2)](z^2 + a^2 + b^2)^{1/2}\right\} \tag{1.146}$$

In the second term of Equation 1.146, make the substitution

$$u^2 = z^2 + a^2 + b^2; \quad z^2 = u^2 - (a^2 + b^2)$$

and obtain

$$ab \int [u^2 - (a^2 + b^2)]^{1/2}[u^2 - (a^2 + b^2) + u^2 - (a^2 + b^2) + (a^2 + b^2)]udu/\{a^2b^2$$

$$+ [u^2 - (a^2 + b^2)]u^2\}u[u^2 - (a^2 + b^2)]^{1/2} \tag{1.147}$$

In Equation 1.147, complete the square of the denominator to obtain

$$\int \{[2abu^2 - ab(a^2 + b^2)]/(u^2 - a^2)(u^2 - b^2)\}du \tag{1.148}$$

Use partial fractions to integrate Equation 1.148. Assume

$$[2abu^2 - ab(a^2 + b^2)]/(u^2 - a^2)(u^2 - b^2)$$

$$= Q/(u + a) + R/(u - a) + S/(u + b) + T/(u - b) \tag{1.149}$$

Expand Equation 1.149 to

$$(u^3 - au^2 - b^2u + ab)Q + (u^3 + au^2 - b^2u - ab^2)R + (u^3 - bu^2 - a^2u + a^2b)$$

$$S + (u^3 + bu^2 - a^2u - a^2b)T = 2abu^2 - ab(a^2 + b^2) \quad (1.150)$$

Equate coefficients

$$u^3 \colon Q + R + S + T = 0 \qquad (1.151)$$

$$u^2 \colon -aQ + aR - bS + bT = 2ab \qquad (1.152)$$

$$u \colon -b^2Q - b2R - a^2 S - a^2 T = 0 \qquad (1.153)$$

$$\text{Constants: } ab^2 Q - ab^2 R + a^2b\, S - a^2bT = -ab(a^2 + b^2) \qquad (1.154)$$

Solve Equations 1.151 to 1.153 simultaneously to yield

$$Q = -b/2; \quad R = b/2; \quad S = -a/2; \quad T = a/2$$

Therefore,

$$O = z \arctan [(z/ab)(z^2 + a^2 + b^2)^{1/2}$$

$$- (b/2)\, Ln[(u + a)/(u - a) + (a/2)Ln[(u + b)/(u - b)] \qquad (1.155)$$

Sum $\{M + N + O + (\pi/2)z\}$ to obtain

$$P_T = (q/2\pi)\{(\pi/2)z - bLn[(u + a)/(u - a)]$$

$$- aLn[(u + b)/(u - b)] - z \arctan[(z/ab)(z^2 + a^2 = b^2)^{1/2} \qquad (1.156)$$

Evaluate Equation 1.154 using limits 0 to h of variable z to obtain

$$P_T = (q/2\pi)\{(\pi/2)h + bLn[(D + a)(C - a)/(D - a)(C + a)]$$

$$+ aLn[(D + b)(C - b)/(D - b)(C + b)] - h \arctan (hC/ab)\} \qquad (1.157)$$

in which

$$D = (a^2 + b^2)^{1/2}$$

Equation 1.157 is the equation Steinbrenner was not able to derive and the equation Terzaghi thought was impossible to solve. To obtain the average vertical stress at a depth h, simply divide Equation 1.157 by h. The average pressure ΔP is used in calculating the footing settlement when underlain by a compressible soil layer.

1.4.1.1 Application

The derived formulas for total vertical stress are applicable in calculating the magnitude of soil settlements due to surface loads and the magnitude of surface loads transmitted to underground structures such as tunnels, bridges, and culverts. The principle of superposition must be utilized in conjunction with the derived formulas to obtain total vertical stresses at corners, center, and any other point of a rectangular area under these surface loads. From these values the total vertical stress on this rectangular area can be calculated to any desired degree of accuracy and, hence, an accurate prediction on the soil settlement and accurate estimate of the transmitted vertical stress on the roof of underground structures can be determined.

Example 1.5

An area 10 ft. (3.05 m) × 20 ft. (6.10 m) is loaded with a uniform load of 1 ksf (47.9 kPa). Tabulate and plot the variation of vertical stress and the total vertical stress through depth at the corner of the rectangular area, using the previously derived formulas for vertical stresses due to a rectangular surface loading.

> **Solution:** Using Microsoft Excel, Table 1.3 shows the tabulated values of the vertical stress, total vertical stress, and average vertical stress through depth in English and system international (SI) units. Figures 1.14, 1.15, and 1.16 are the plots of the vertical stress distribution in the English units and Figures 1.17, 1.18, and 1.19 are the plots of the vertical stress distribution in the SI units.

It should be noted that once the Excel Program has been set up, it is only a matter of changing the parameters of a, b, and q to solve for vertical stresses for any other problem involving a uniform surface load on a rectangular area.

The method of superposition is applied to other cases in which the point under consideration is within, outside, and along the boundaries of the loaded area. This is shown not only for determining transmitted pressure to roofs of underground structures but also to determine the amount of settlements due to surface loads. These surface loads represent not only vehicular loading but also pressures from footing foundations, which can contribute to the magnitude of soil settlement underneath.

The reader is assumed to be able to substitute these numerical values in established formulas for soil settlements due to these pressures. The standard settlement formula is given in textbooks as

$$S = [C_c/(1 + e_o)]\log_{10}\{(p_o + \Delta p)/p_o\}H \tag{1.158}$$

in which C_c = compression index, H = the thickness of the compressible soil layer, p_o = the pressure due to soil from the surface up to the center of the soil layer of thickness H underneath the foundation, e_o = the initial void ratio of the compressible soil layer, and Δp = additional pressure due to the footing foundation above the soil layer.

Table 1.3 Vertical Stress Values

	ENGLISH UNITS				SI UNITS		
h in ft.	P in psf	PT in lb/ft.	Ave P in psf	h in m	P in kPa	PT in kN/m	Avg P in kPa
1	250	250	250	0.30	12	4	12
2	249	500	250	0.61	12	7	12
3	247	748	249	0.91	12	11	12
4	244	994	248	1.22	12	14	12
5	239	1235	247	1.52	11	18	12
6	233	1471	245	1.83	11	21	12
7	226	1701	243	2.13	11	25	12
8	218	1922	240	2.44	10	28	12
9	209	2136	237	2.74	10	31	11
10	200	2340	234	3.05	10	34	11
11	191	2536	231	3.35	9	37	11
12	182	2722	227	3.66	9	40	11
13	173	2899	223	3.96	8	42	11
14	164	3068	219	4.27	8	45	10
15	156	3228	215	4.57	7	47	10
16	148	3380	211	4.88	7	49	10
17	141	3525	207	5.18	7	51	10
18	133	3662	203	5.49	6	53	10
19	127	3792	200	5.79	6	55	10
20	120	3915	196	6.10	6	57	9
21	114	4032	192	6.40	5	59	9
22	108	4143	188	6.71	5	60	9
23	103	4249	185	7.01	5	62	9
24	98	4349	181	7.32	5	63	9
25	93	4445	178	7.62	4	65	9
26	89	4536	174	7.92	4	66	8
27	84	4622	171	8.23	4	67	8
28	80	4705	168	8.53	4	69	8
29	77	4783	165	8.84	4	70	8
30	73	4858	162	9.14	4	71	8
31	70	4930	159	9.45	3	72	8
32	67	4998	156	9.75	3	73	7
33	64	5063	153	10.06	3	74	7
34	61	5126	151	10.36	3	75	7
35	59	5186	148	10.67	3	76	7
36	56	5243	146	10.97	3	77	7
37	54	5298	143	11.28	3	77	7
38	52	5351	141	11.58	2	78	7
39	49	5401	138	11.89	2	79	7
40	48	5450	136	12.19	2	80	7
41	46	5496	134	12.50	2	80	6
42	44	5541	132	12.80	2	81	6
43	42	5584	130	13.11	2	81	6
44	41	5626	128	13.41	2	82	6
45	39	5666	126	13.72	2	83	6
46	38	5704	124	14.02	2	83	6
47	36	5741	122	14.33	2	84	6
48	35	5777	120	14.63	2	84	6
49	34	5811	119	14.94	2	85	6
50	33	5845	117	15.24	2	85	6
51	32	5877	115	15.54	2	86	6
52	31	5908	114	15.85	1	86	5
53	30	5938	112	16.15	1	87	5
54	29	5968	111	16.46	1	87	5
55	28	5996	109	16.76	1	87	5
56	27	6023	108	17.07	1	88	5
57	26	6050	106	17.37	1	88	5
58	25	6075	105	17.68	1	89	5
59	25	6100	103	17.98	1	89	5
60	24	6124	102	18.29	1	89	5
61	23	6148	101	18.59	1	90	5
62	22	6171	100	18.90	1	90	5
63	22	6193	98	19.20	1	90	5
64	21	6214	97	19.51	1	91	5
65	21	6235	96	19.81	1	91	5
66	20	6255	95	20.12	1	91	5
67	19	6275	94	20.42	1	92	4
68	19	6294	93	20.73	1	92	4
69	18	6313	91	21.03	1	92	4
70	18	6331	90	21.34	1	92	4

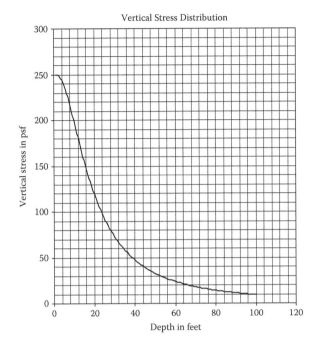

Figure 1.14 Vertical stress distribution.

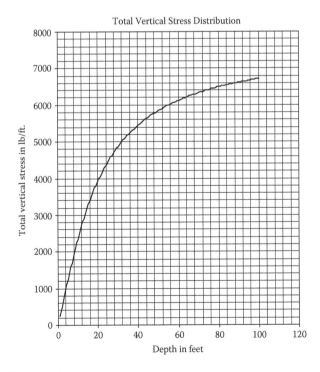

Figure 1.15 Total vertical stress distribution.

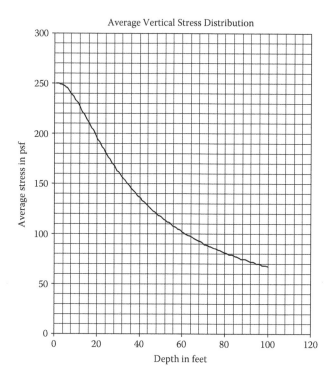

Figure 1.16 Average vertical stress distribution.

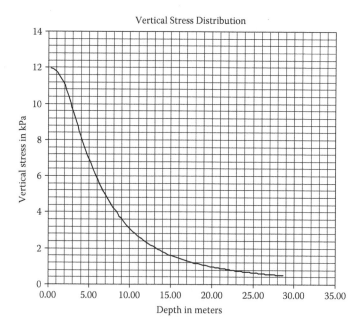

Figure 1.17 Vertical stress distribution.

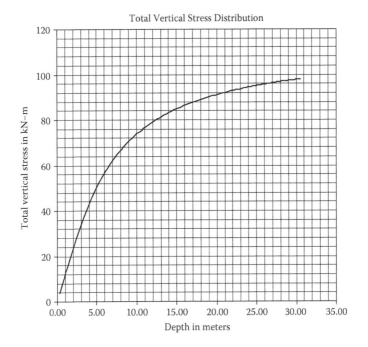

Figure 1.18 Total vertical stress distribution.

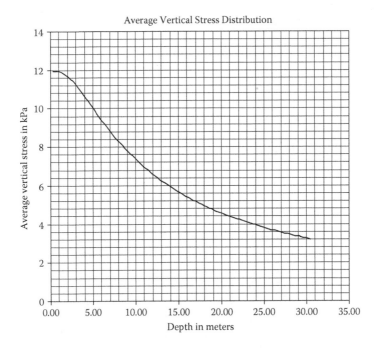

Figure 1.19 Average vertical stress distribution.

The calculation of Δp has been done by using Newmark's chart, Steinbrenner's formula given by Equation 1.138 or 1.139, or finite element calculation methods.

In our case we shall be using the exact method of Equation 1.159, by applying it first up to the bottom of the soil layer and then at the top of the soil layer. Obtain the difference of the total pressures and divide this value by H to obtain the exact Δp to use in Equation 1.158. Since the above equations for the dispersion of surface loads through soil are generic to English and SI units, we can chose whatever unit we are familiar with.

Example 1.6

In the above example, determine the vertical stress values and plots of the vertical stress, total vertical stress, and average vertical stress through depth when the point under consideration is under the center of the rectangular area.

> **Solution:** Using Excel, Table 1.4 shows the tabulated values of the vertical stress, total vertical stress, and average vertical stress through depth, and Figures 1.20, 1.21, and 1.22 are the plots of the vertical stress distributions in English units, while Figures 1.23, 1.24, and 1.25 are the plots in the SI units.

For Example 1.6, the derived formulas are applied to the divided area using the principle of superposition (in effect, four rectangular areas are calculated) to obtain the values for a point under the center of the given rectangular area.

Note that we are showing the graph of the pressure distribution just to picture the distribution of the surface load through the soil medium. In practice we can directly apply Equation 1.157 to any specific depth in a problem.

Most of the following examples are in the English system of units, and that is how they are set up by the author. For SI units we need to apply the conversion formulas to relate the English to the SI units of measurements.

We shall investigate the different cases where we can apply Equations 1.157 and 1.158 to solve settlement problems at any location relative to our footing foundation. The following illustrations will show four cases. The author believes these cases are sufficient to solve any settlement problems involving surface loads and compressible soil layers underneath them.

Case 1: Any point along the perimeter of a loaded area. Figure 1.26 illustrates this condition. At any point along the perimeter of the loaded area, the vertical stress at a depth under point A is given by the sum of the vertical stresses due to the rectangles ABCD and ADEO.

The value of b in the formula for vertical stress is $(b - y)$ for rectangle ABCD and y for rectangle ADEO. The value of a in the formula for vertical stress for both rectangles is the same. Since dimensions a and b are interchangeable, y can be replaced by x to complete the analysis along the perimeter of the loaded area.

Table 1.4 Vertical Stress Values

		English Units		h in m		SI Units	
h in ft.	P in psf	PT in lb/ft.	Ave P in psf	h in m	P in kPa	PT in kN/m	Avg P in kPa
1	997	999	999	0.30	48	15	48
2	976	1987	994	0.61	47	29	48
3	932	2943	981	0.91	45	43	47
4	870	3845	961	1.22	42	56	46
5	800	4680	936	1.52	38	68	45
6	727	5444	907	1.83	35	79	43
7	658	6136	877	2.13	31	90	42
8	593	6761	845	2.44	28	99	40
9	534	7323	814	2.74	26	107	39
10	481	7830	783	3.05	23	114	37
11	433	8287	753	3.35	21	121	36
12	392	8699	725	3.66	19	127	35
13	355	9072	698	3.96	17	132	33
14	322	9409	672	4.27	15	137	32
15	293	9717	648	4.57	14	142	31
16	267	9996	625	4.88	13	146	30
17	244	10252	603	5.18	12	150	29
18	224	10486	583	5.49	11	153	28
19	206	10701	563	5.79	10	156	27
20	190	10899	545	6.10	9	159	26
21	176	11082	528	6.40	8	162	25
22	163	11251	511	6.71	8	164	24
23	151	11408	496	7.01	7	166	24
24	141	11554	481	7.32	7	169	23
25	131	11690	468	7.62	6	171	22
26	123	11817	454	7.92	6	172	22
27	115	11935	442	8.23	5	174	21
28	108	12046	430	8.53	5	176	21
29	101	12151	419	8.84	5	177	20
30	95	12249	408	9.14	5	179	20
31	90	12341	398	9.45	4	180	19
32	85	12428	388	9.75	4	181	19
33	80	12511	379	10.06	4	183	18
34	76	12589	370	10.36	4	184	18
35	72	12663	362	10.67	3	185	17
36	68	12733	354	10.97	3	186	17
37	65	12799	346	11.28	3	187	17
38	62	12862	338	11.58	3	188	16
39	59	12923	331	11.89	3	189	16
40	56	12980	324	12.19	3	189	16
41	54	13035	318	12.50	3	190	15
42	51	13087	312	12.80	2	191	15
43	49	13137	306	13.11	2	192	15
44	47	13185	300	13.41	2	192	14
45	45	13231	294	13.72	2	193	14
46	43	13275	289	14.02	2	194	14
47	41	13317	283	14.33	2	194	14
48	40	13357	278	14.63	2	195	13
49	38	13396	273	14.94	2	195	13
50	37	13434	269	15.24	2	196	13
51	35	13470	264	15.54	2	197	13
52	34	13504	260	15.85	2	197	12
53	33	13538	255	16.15	2	198	12
54	32	13570	251	16.46	2	198	12
55	31	13601	247	16.76	1	198	12
56	29	13631	243	17.07	1	199	12
57	28	13660	240	17.37	1	199	11
58	28	13688	236	17.68	1	200	11
59	27	13715	232	17.98	1	200	11
60	26	13741	229	18.29	1	201	11
61	25	13766	226	18.59	1	201	11
62	24	13791	222	18.90	1	201	11
63	23	13815	219	19.20	1	202	10
64	23	13838	216	19.51	1	202	10
65	22	13860	213	19.81	1	202	10
66	21	13882	210	20.12	1	203	10
67	21	13903	208	20.42	1	203	10
68	20	13924	205	20.73	1	203	10
69	20	13944	202	21.03	1	203	10
70	19	13963	199	21.34	1	204	10

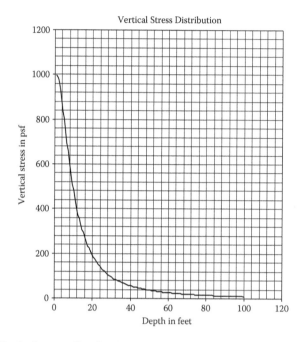

Figure 1.20 Vertical stress distribution.

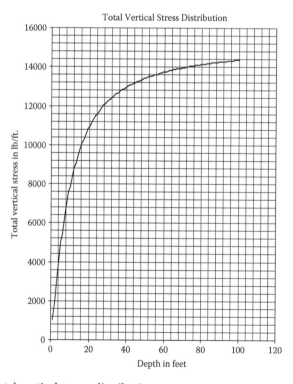

Figure 1.21 Total vertical stress distribution.

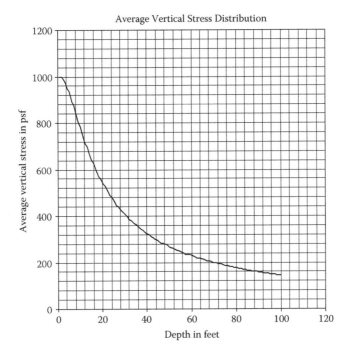

Figure 1.22 Average vertical stress distribution.

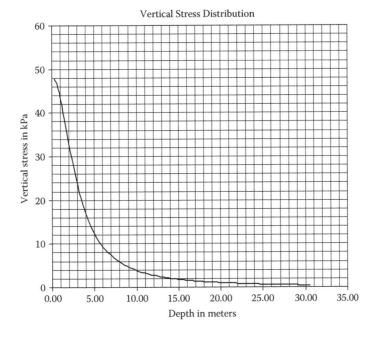

Figure 1.23 Vertical stress distribution.

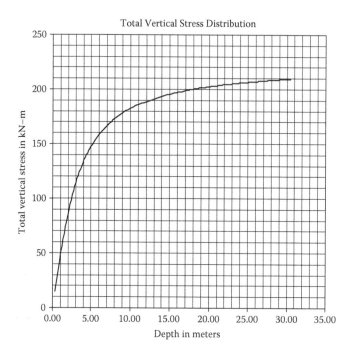

Figure 1.24 Total vertical stress distribution.

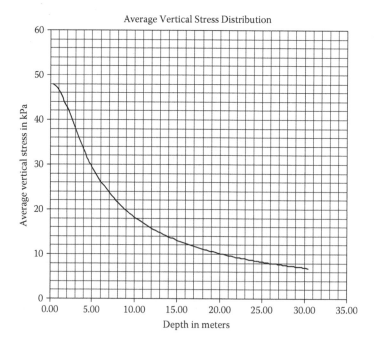

Figure 1.25 Average vertical stress distribution.

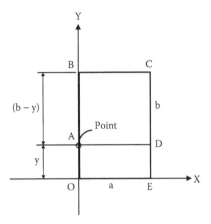

Figure 1.26 Point along perimeter of loaded area.

Example 1.7

Calculate the vertical stresses under point A when y = 5 ft. in Figure 1.24. Use English units from here on.

> **Solution:** Using the same Excel program as in Example 1.2, use rectangles ABCD and DEO, whose dimensions are 15 × 10 ft. and 5 × 10 ft., respectively, and add the results to obtain the vertical stresses under the point. Table 1.5 and Figures 1.27 and 1.28 are the required solutions to this problem.

Case 2: Any point inside the loaded area. Figure 1.29 illustrates this condition. At any point inside the loaded area the vertical stress is obtained by

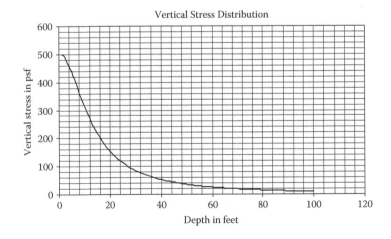

Figure 1.27 Vertical stress distribution.

Table 1.5 Tabulated Values of Vertical Stresses

	Vertical Stress Values				Total Vertical Stress Values			
h	ABCD	ADEO	h in ft.	P in psf	ABCD	ADEO	h in ft.	Pₒₜ
1	250	249	1	499	250	250	1	500
2	249	244	2	493	500	497	2	996
3	247	233	3	480	748	736	3	1483
4	243	218	4	461	993	961	4	1954
5	238	200	5	438	1233	1170	5	2404
6	231	182	6	413	1468	1361	6	2829
7	223	164	7	387	1695	1534	7	3229
8	214	148	8	362	1913	1690	8	3603
9	204	133	9	337	2122	1831	9	3953
10	194	120	10	314	2320	1958	10	4278
11	183	108	11	292	2509	2072	11	4581
12	173	98	12	271	2687	2175	12	4862
13	164	89	13	252	2856	2268	13	5124
14	154	80	14	235	3015	2352	14	5367
15	145	73	15	218	3164	2429	15	5593
16	136	67	16	203	3305	2499	16	5804
17	128	61	17	190	3437	2563	17	6000
18	121	56	18	177	3562	2622	18	6183
19	114	52	19	165	3679	2675	19	6354
20	107	48	20	155	3789	2725	20	6514
21	101	44	21	145	3893	2770	21	6664
22	95	41	22	136	3991	2813	22	6804
23	90	38	23	128	4084	2852	23	6936
24	85	35	24	120	4171	2888	24	7059
25	80	33	25	113	4253	2922	25	7176
26	76	31	26	106	4331	2954	26	7285
27	72	29	27	100	4405	2984	27	7389
28	68	27	28	95	4475	3012	28	7486
29	64	25	29	90	4541	3038	29	7579
30	61	24	30	85	4604	3062	30	7666
31	58	22	31	81	4664	3085	31	7749
32	55	21	32	77	4720	3107	32	7827
33	53	20	33	73	4774	3128	33	7902
34	50	19	34	69	4826	3147	34	7973
35	48	18	35	66	4875	3166	35	8040
36	46	17	36	63	4922	3183	36	8105
37	44	16	37	60	4966	3200	37	8166
38	42	15	38	57	5009	3216	38	8225
39	40	15	39	55	5050	3231	39	8281
40	38	14	40	52	5089	3245	40	8334
41	37	13	41	50	5127	3259	41	8385
42	35	13	42	48	5163	3272	42	8434
43	34	12	43	46	5197	3284	43	8481
44	32	12	44	44	5230	3296	44	8526
45	31	11	45	42	5262	3308	45	8570
46	30	11	46	41	5293	3319	46	8611
47	29	10	47	39	5322	3329	47	8651
48	28	10	48	38	5350	3339	48	8690
49	27	10	49	36	5378	3349	49	8727
50	26	9	50	35	5404	3358	50	8762
51	25	9	51	34	5429	3367	51	8797
52	24	9	52	33	5454	3376	52	8830
53	23	8	53	31	5478	3384	53	8862
54	22	8	54	30	5501	3392	54	8893
55	22	8	55	29	5523	3400	55	8923
56	21	7	56	28	5544	3408	56	8952
57	20	7	57	27	5565	3415	57	8980
58	20	7	58	27	5585	3422	58	9007
59	19	7	59	26	5604	3429	59	9033
60	19	6	60	25	5623	3435	60	9058
61	18	6	61	24	5641	3442	61	9083
62	17	6	62	23	5659	3448	62	9107
63	17	6	63	23	5676	3454	63	9130
64	16	6	64	22	5693	3459	64	9152
65	16	6	65	21	5709	3465	65	9174
66	15	5	66	21	5724	3471	66	9195
67	15	5	67	20	5740	3476	67	9216
68	15	5	68	20	5755	3481	68	9235
69	14	5	69	19	5769	3486	69	9255
70	14	5	70	19	5783	3491	70	9274
71	13	5	71	18	5797	3495	71	9292

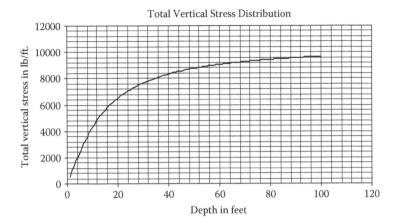

Figure 1.28 Total vertical stress distribution.

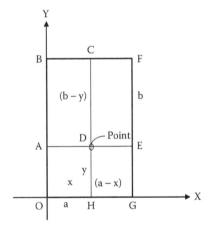

Rectangle	Dimensions	
	a	b
ABCD	x	(b − y)
CDEF	(a − x)	(b − y)
ADHO	x	y
DEGH	(a − x)	y

Figure 1.29 Point inside the loaded area.

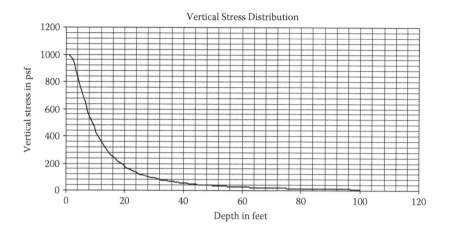

Figure 1.30 Vertical stress distribution.

the sum of vertical stresses due to rectangles ABCD, CDEF, ADHO, and DEGH. Dimensions for a and b in the vertical stress formula are as follows:

Example 1.8

Using the same rectangular area loaded with a uniform loading of 1 ksf, determine the vertical stresses underneath the point located by $x = 4$ ft. and $y = 6$ ft.

> **Solution:** Using the excel program already set up, tabulate values and plot as follows: Table 1.6 shows the tabulated values of vertical stresses. Figures 1.30 and 1.31 are the plots of the vertical stress and total vertical stress, respectively.

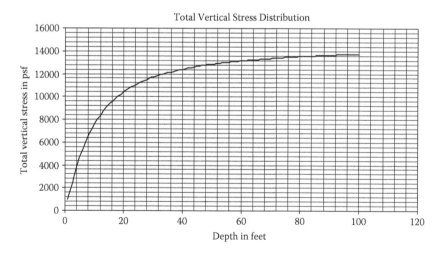

Figure 1.31 Total vertical stress distribution.

Table 1.6 Tabulated Values of Vertical Stresses

Vertical Stress Values						Total Vertical Stress Values					
ABCD	CDEF	DEGH	ADHO	h in ft.	P in psf	ABCD	CDEF	DEGH	ADHO	h in ft.	PT in lb/ft.
248	250	249	248	1	995	250	250	250	250	1	999
240	246	244	238	2	968	494	498	497	493	2	1983
224	239	232	218	3	914	727	741	736	722	3	2926
204	229	216	194	4	842	941	976	960	928	4	3805
184	216	196	168	5	764	1134	1198	1166	1109	5	4608
165	202	175	145	6	687	1309	1407	1351	1266	6	5333
148	187	155	125	7	616	1465	1602	1517	1400	7	5984
134	173	137	107	8	551	1606	1782	1663	1516	8	6567
121	160	121	92	9	494	1733	1949	1792	1615	9	7089
109	147	107	80	10	444	1848	2102	1906	1701	10	7557
99	136	95	70	11	400	1952	2244	2006	1776	11	7979
91	125	84	61	12	361	2047	2375	2095	1842	12	8359
83	116	75	54	13	328	2134	2495	2175	1899	13	8703
76	107	67	48	14	298	2213	2606	2246	1950	14	9015
70	99	60	43	15	272	2286	2709	2309	1995	15	9300
64	92	54	38	16	249	2353	2805	2366	2036	16	9560
59	85	49	34	17	228	2415	2893	2418	2072	17	9798
55	79	45	31	18	210	2472	2975	2465	2105	18	10017
51	74	41	28	19	194	2525	3051	2508	2135	19	10219
47	69	37	26	20	179	2574	3122	2547	2162	20	10405
44	64	34	24	21	166	2620	3189	2583	2186	21	10578
41	60	32	22	22	154	2663	3251	2615	2209	22	10738
38	56	29	20	23	144	2702	3309	2646	2230	23	10887
36	53	27	19	24	134	2740	3363	2674	2249	24	11025
34	49	25	17	25	125	2774	3414	2700	2267	25	11155
32	46	23	16	26	117	2807	3462	2724	2284	26	11276
30	44	22	15	27	110	2838	3507	2747	2299	27	11390
28	41	20	14	28	103	2867	3549	2768	2313	28	11497
26	39	19	13	29	97	2894	3589	2787	2327	29	11597
25	37	18	12	30	92	2919	3627	2806	2339	30	11692
24	35	17	11	31	87	2944	3663	2823	2351	31	11781
22	33	16	11	32	82	2967	3697	2840	2362	32	11865
21	31	15	10	33	78	2989	3729	2855	2372	33	11945
20	30	14	10	34	74	3009	3760	2870	2382	34	12021
19	28	13	9	35	70	3029	3789	2883	2392	35	12092
18	27	13	9	36	66	3047	3816	2896	2400	36	12161
17	26	12	8	37	63	3065	3843	2909	2409	37	12225
17	25	11	8	38	60	3082	3868	2920	2417	38	12287
16	23	11	7	39	57	3098	3892	2932	2424	39	12346
15	22	10	7	40	55	3114	3915	2942	2431	40	12402
14	21	10	7	41	52	3129	3937	2952	2438	41	12456
14	21	9	6	42	50	3143	3958	2962	2445	42	12507
13	20	9	6	43	48	3156	3978	2971	2451	43	12556
13	19	9	6	44	46	3169	3997	2980	2457	44	12603
12	18	8	6	45	44	3181	4015	2988	2462	45	12648
12	17	8	5	46	42	3193	4033	2996	2468	46	12691
11	17	8	5	47	41	3205	4050	3004	2473	47	12732
11	16	7	5	48	39	3216	4067	3012	2478	48	12772
10	15	7	5	49	38	3226	4082	3019	2483	49	12810
10	15	7	5	50	36	3237	4098	3026	2487	50	12847
10	14	6	4	51	35	3246	4112	3032	2492	51	12883
9	14	6	4	52	34	3256	4126	3038	2496	52	12917
9	13	6	4	53	32	3265	4140	3045	2500	53	12950
9	13	6	4	54	31	3274	4153	3050	2504	54	12981
8	12	6	4	55	30	3282	4166	3056	2508	55	13012
8	12	5	4	56	29	3291	4178	3062	2511	56	13042
8	12	5	3	57	28	3298	4190	3067	2515	57	13070
8	11	5	3	58	27	3306	4201	3072	2518	58	13098
7	11	5	3	59	26	3314	4212	3077	2522	59	13125
7	11	5	3	60	26	3321	4223	3082	2525	60	13151
7	10	5	3	61	25	3328	4234	3086	2528	61	13176
7	10	4	3	62	24	3335	4244	3091	2531	62	13200
6	10	4	3	63	23	3341	4253	3095	2534	63	13224
6	9	4	3	64	23	3347	4263	3099	2537	64	13247
6	9	4	3	65	22	3354	4272	3103	2540	65	13269
6	9	4	3	66	21	3360	4281	3107	2542	66	13290
6	9	4	3	67	21	3365	4290	3111	2545	67	13311
6	8	4	2	68	20	3371	4298	3115	2547	68	13331
5	8	4	2	69	19	3377	4306	3119	2550	69	13351
5	8	3	2	70	19	3382	4314	3122	2552	70	13370
5	8	3	2	71	18	3387	4322	3126	2554	71	13389

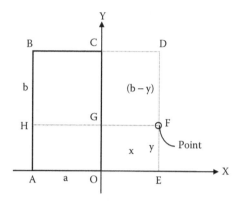

Rectangle	Dimensions	
	a	b
BDFH	(a + x)	(b − y)
CDFG	x	(b − y)
AEFH	(a + x)	y
EFGO	x	y

Figure 1.32 Point outside the loaded area.

Case 3: Any point outside the loaded area. Figure 1.32 illustrates this condition. The vertical stress at any point outside the perimeter of the loaded area will involve addition and subtraction of rectangular areas BDFH, AEFH, CDFG, and EFGO such that the value of the expression

$$BDFH + AEFH - (CDFG + EFGO) \qquad (1.159)$$

will represent the vertical stress at a depth under point F due to loaded area ABCO.

Example 1.9

Determine the values and plots of the vertical stresses underneath the point when x = 8 ft. and y = 9 ft. Use the same rectangular area and intensity of 1 ksf uniform load.

> **Solution:** Table 1.7 and Figures 1.33 and 1.34 are the required solutions from Excel. Note the changes in the shape of the plots of the vertical stresses when the point is outside the rectangular area.

Table 1.7 Tabulated Values of Vertical Stresses

	Vertical Stress Values						Total Vertical Stress Values				
BDFH	AEFH	CDFG	EFGO	h in ft.	P in psf	BDFH	AEFH	CDFG	EFGO	h in ft.	P_T in lb/ft.
250	250	250	250	1	0	250	250	250	250	1	0
249	249	248	248	2	2	500	499	499	499	2	1
248	246	244	243	3	7	748	747	745	744	3	6
245	242	237	235	4	15	995	991	986	983	4	17
241	236	228	224	5	25	1238	1230	1219	1213	5	37
236	228	216	211	6	37	1476	1463	1441	1430	6	68
229	219	204	197	7	48	1708	1687	1651	1634	7	110
222	210	191	182	8	59	1934	1901	1849	1823	8	164
213	200	178	167	9	68	2151	2106	2033	1998	9	227
205	190	165	154	10	76	2361	2301	2204	2158	10	300
196	180	152	141	11	83	2561	2486	2362	2305	11	380
187	170	141	129	12	88	2753	2661	2508	2440	12	465
178	161	130	118	13	91	2935	2826	2644	2563	13	555
170	152	120	108	14	93	3109	2982	2768	2676	14	647
161	143	111	99	15	94	3274	3130	2884	2780	15	741
153	135	102	91	16	95	3431	3269	2990	2875	16	835
145	127	95	84	17	94	3580	3400	3089	2962	17	930
138	120	88	77	18	93	3722	3524	3180	3042	18	1023
131	113	82	71	19	91	3856	3640	3265	3116	19	1115
124	107	76	66	20	89	3983	3751	3343	3185	20	1205
117	101	71	61	21	87	4104	3855	3417	3248	21	1293
111	96	66	57	22	85	4218	3953	3485	3307	22	1379
106	91	62	53	23	82	4327	4046	3548	3362	23	1462
100	86	58	49	24	79	4430	4135	3608	3413	24	1543
95	81	54	46	25	77	4528	4218	3664	3461	25	1621
91	77	51	43	26	74	4621	4297	3716	3505	26	1697
86	73	48	40	27	72	4709	4372	3765	3547	27	1770
82	70	45	38	28	69	4793	4444	3811	3586	28	1840
78	66	42	36	29	66	4874	4512	3855	3623	29	1908
75	63	40	34	30	64	4950	4576	3896	3658	30	1973
71	60	38	32	31	62	5023	4638	3934	3690	31	2036
68	57	36	30	32	59	5093	4696	3971	3721	32	2096
65	54	34	28	33	57	5159	4752	4006	3751	33	2155
62	52	32	27	34	55	5223	4805	4039	3778	34	2211
59	50	30	26	35	53	5283	4856	4070	3804	35	2265
57	48	29	24	36	51	5341	4905	4100	3829	36	2317
54	45	28	23	37	49	5397	4951	4128	3853	37	2367
52	44	26	22	38	47	5450	4996	4155	3876	38	2415
50	42	25	21	39	46	5501	5038	4181	3897	39	2462
48	40	24	20	40	44	5550	5079	4205	3917	40	2507
46	38	23	19	41	43	5597	5118	4229	3937	41	2550
44	37	22	18	42	41	5643	5156	4251	3956	42	2592
43	35	21	17	43	40	5686	5192	4272	3973	43	2633
41	34	20	17	44	38	5728	5227	4293	3991	44	2672
39	33	19	16	45	37	5768	5260	4313	4007	45	2709
38	32	19	15	46	36	5807	5293	4331	4023	46	2746
37	30	18	15	47	35	5844	5324	4350	4038	47	2781
35	29	17	14	48	33	5880	5353	4367	4052	48	2815
34	28	16	14	49	32	5915	5382	4384	4066	49	2848
33	27	16	13	50	31	5949	5410	4400	4079	50	2879
32	26	15	13	51	30	5981	5437	4415	4092	51	2910
31	25	15	12	52	29	6012	5463	4430	4105	52	2940
30	25	14	12	53	28	6043	5488	4445	4117	53	2969
29	24	14	11	54	28	6072	5512	4459	4128	54	2997
28	23	13	11	55	27	6100	5535	4472	4139	55	3024
27	22	13	11	56	26	6128	5558	4485	4150	56	3050
26	22	12	10	57	25	6154	5580	4498	4160	57	3076
25	21	12	10	58	24	6180	5601	4510	4170	58	3101
25	20	12	10	59	24	6205	5622	4522	4180	59	3125
24	20	11	9	60	23	6229	5642	4533	4189	60	3148
23	19	11	9	61	22	6252	5661	4544	4199	61	3171
22	19	11	9	62	22	6275	5680	4555	4207	62	3193
22	18	10	8	63	21	6297	5698	4565	4216	63	3214
21	17	10	8	64	21	6319	5716	4575	4224	64	3235
21	17	10	8	65	20	6340	5733	4585	4232	65	3256
20	16	9	8	66	19	6360	5750	4594	4240	66	3275
19	16	9	7	67	19	6380	5766	4604	4248	67	3295
19	16	9	7	68	18	6399	5782	4612	4255	68	3313
18	15	9	7	69	18	6418	5797	4621	4262	69	3332
18	15	8	7	70	18	6436	5812	4630	4269	70	3349
17	14	8	7	71	17	6453	5827	4638	4276	71	3367

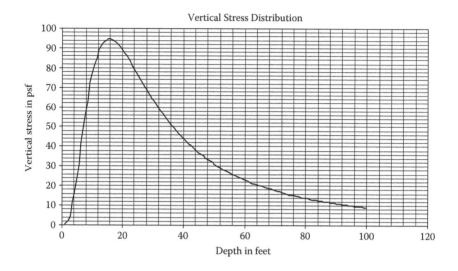

Figure 1.33 Vertical stress distribution.

Case 4: Any point distant from adjacent sides. In Figure 1.35 the vertical stress at a point distant from adjacent sides and outside the loaded area involves the relationships of rectangles ACEO, ABFO, DEOH, and FGHO such that the relation

$$ACDO + EGHO - (ABEO + DFHO) \qquad (1.160)$$

will determine the value of the vertical stress under point O.

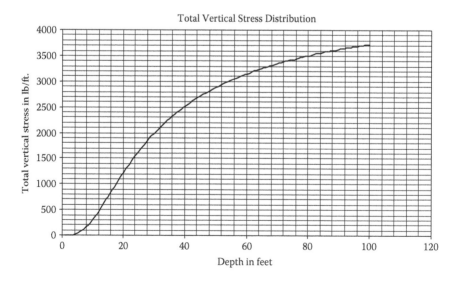

Figure 1.34 Total vertical stress distribution.

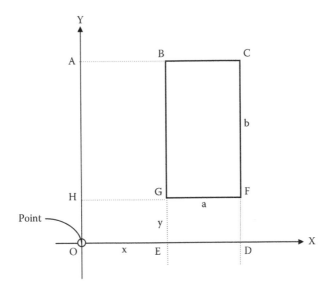

Rectangle	Dimensions	
	a	b
ACDO	(a + x)	(b + y)
ABEO	x	(b + y)
DFHO	(a + x)	y
EGHO	x	y

Figure 1.35 Point distant from adjacent sides.

Example 1.10

Determine the values and plot of the vertical stresses under the point when x = 11 ft. and y = 7 ft.

Solution: Table 1.8 and Figures 1.36 and 1.37 are the solutions from Excel.

1.4.2   Triangular load on a rectangular area

For more than half a century, the integration of Boussinesq's elastic equation for a triangular surface loading has eluded previous investigators on this subject. As a result, graphical and finite-element procedures are currently being utilized to solve this problem. The following derivations show the results of the integration of the elastic equation for total vertical stress due

Table 1.8 Tabulated Values of Vertical Stresses

		Vertical Stress Values						Total Vertical Stress Values			
ACDO	EGHO	ABEO	DFHO	h in ft.	P in psf	ACDO	EGHO	ABEO	DFHO	h in ft.	PT in lb/ft.
250	250	250	250	1	0	250	250	250	250	1	0
250	247	249	248	2	0	500	499	500	499	2	0
250	242	248	243	3	1	750	744	748	744	3	0
249	233	245	236	4	1	999	982	995	984	4	1
248	222	242	226	5	2	1248	1210	1239	1215	5	3
247	209	237	215	6	4	1495	1425	1478	1436	6	6
245	195	231	203	7	6	1742	1627	1713	1645	7	11
243	181	225	192	8	8	1986	1815	1941	1843	8	18
241	167	217	180	9	10	2228	1989	2162	2028	9	27
238	154	210	169	10	13	2468	2149	2375	2203	10	39
235	141	202	159	11	15	2704	2296	2581	2367	11	52
231	130	194	149	12	17	2937	2432	2780	2520	12	69
227	119	187	140	13	20	3166	2556	2970	2665	13	87
223	109	179	131	14	22	3391	2670	3153	2800	14	108
218	101	172	124	15	24	3612	2775	3329	2928	15	131
214	93	164	116	16	26	3828	2872	3497	3047	16	156
209	86	157	109	17	27	4039	2961	3657	3160	17	182
204	79	151	103	18	29	4245	3043	3812	3267	18	210
199	73	144	97	19	30	4447	3119	3959	3367	19	240
193	68	138	92	20	31	4643	3190	4101	3462	20	270
188	63	132	87	21	32	4833	3256	4236	3551	21	302
183	59	127	82	22	33	5019	3317	4366	3636	22	334
178	55	122	78	23	33	5199	3374	4490	3716	23	367
173	51	117	74	24	33	5374	3427	4609	3792	24	400
168	48	112	70	25	34	5544	3476	4723	3864	25	434
163	45	107	67	26	34	5709	3523	4832	3932	26	467
158	42	103	63	27	34	5869	3566	4937	3997	27	501
153	40	99	60	28	34	6025	3607	5038	4059	28	535
148	37	95	57	29	33	6175	3646	5135	4118	29	568
144	35	91	55	30	33	6321	3682	5227	4174	30	602
139	33	87	52	31	33	6462	3717	5317	4227	31	635
135	32	84	50	32	33	6599	3749	5402	4278	32	668
131	30	81	48	33	32	6732	3780	5485	4327	33	700
127	28	78	45	34	32	6860	3809	5564	4373	34	732
123	27	75	43	35	31	6985	3836	5640	4418	35	763
119	26	72	42	36	31	7106	3863	5714	4460	36	794
115	24	69	40	37	30	7223	3888	5785	4501	37	825
112	23	67	38	38	30	7336	3911	5853	4540	38	854
108	22	65	37	39	29	7446	3934	5919	4578	39	884
105	21	62	35	40	29	7553	3956	5982	4613	40	913
102	20	60	34	41	28	7656	3976	6043	4648	41	941
99	19	58	33	42	27	7756	3996	6102	4681	42	969
96	18	56	31	43	27	7853	4015	6159	4713	43	996
93	18	54	30	44	26	7948	4033	6215	4744	44	1022
90	17	52	29	45	26	8039	4050	6268	4773	45	1048
87	16	51	28	46	25	8128	4067	6319	4802	46	1074
85	16	49	27	47	25	8214	4083	6369	4829	47	1099
82	15	47	26	48	24	8298	4098	6417	4856	48	1123
80	14	46	25	49	24	8379	4113	6464	4881	49	1147
78	14	45	24	50	23	8458	4127	6509	4906	50	1170
76	13	43	23	51	23	8535	4141	6553	4930	51	1193
74	13	42	23	52	22	8609	4154	6596	4953	52	1215
71	12	41	22	53	22	8682	4167	6637	4975	53	1237
70	12	39	21	54	21	8752	4179	6677	4996	54	1258
68	12	38	20	55	21	8821	4191	6715	5017	55	1279
66	11	37	20	56	20	8888	4202	6753	5037	56	1299
64	11	36	19	57	20	8953	4213	6790	5057	57	1319
62	10	35	19	58	19	9016	4224	6825	5076	58	1339
61	10	34	18	59	19	9077	4234	6860	5094	59	1358
59	10	33	18	60	18	9137	4244	6893	5112	60	1376
58	10	32	17	61	18	9196	4254	6926	5129	61	1395
56	9	31	17	62	18	9253	4263	6957	5146	62	1412
55	9	30	16	63	17	9308	4272	6988	5162	63	1430
53	9	30	16	64	17	9362	4281	7018	5178	64	1447
52	8	29	15	65	16	9415	4290	7047	5194	65	1464
51	8	28	15	66	16	9466	4298	7076	5208	66	1480
50	8	27	14	67	16	9516	4306	7104	5223	67	1496
48	8	27	14	68	15	9565	4314	7131	5237	68	1511
47	7	26	14	69	15	9613	4321	7157	5251	69	1527
46	7	25	13	70	15	9660	4329	7183	5264	70	1542
45	7	25	13	71	14	9705	4336	7208	5277	71	1556

to a triangular loading using standard integration formulas. Three levels of substitutions have been employed to solve these problems.

1.4.2.1 Derivation

The vertical stress under the corner of a rectangular area due to a uniform line load is given by the equation

$$P = \{qah^3/2\pi\}[\{(2a^2 + 3h^2)\,b + 3y^2\}/(h^2 + y^2)^2\,(h^2 + a^2 + y^2)^{3/2}] \quad (1.161)$$

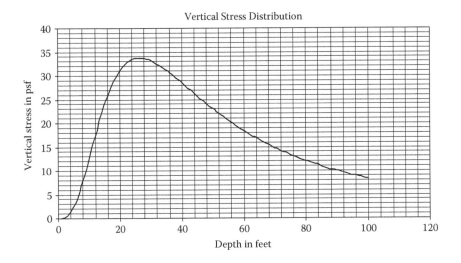

Figure 1.36 Vertical stress distribution.

In Figure 1.38, the vertical stress under the zero load for a rectangular area loaded with a triangular load can be expressed by

$$P_o = (qah^3/2\pi b) \int_0^b [\{2a^2 + 3h^2\}y + 3y^3\}/(h^2 + y^2)^2 (h^2 + a^2 + y^2)^{3/2}]dy \quad (1.162)$$

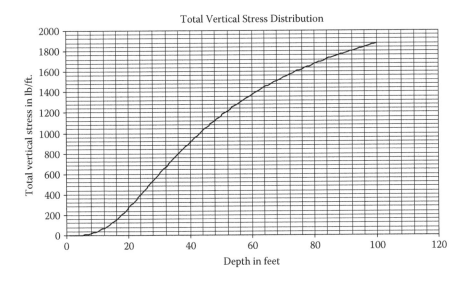

Figure 1.37 Total vertical stress distribution.

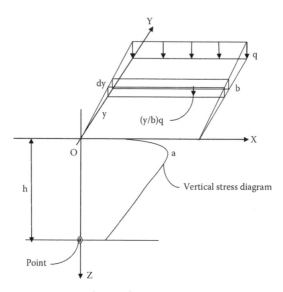

Figure 1.38 Vertical stress under a triangular load intensity.

Let

$$I = (2a^2 + 3h^2) \int_0^b [y/(h^2 + y^2)^2 \, (h^2 + a^2 + y^2)^{3/2}] dy \qquad (1.163)$$

$$II = 3 \int_0^b [y^3/(h^2 + y^2)^2 \, (h^2 + a^2 + y^2)^{3/2}] dy \qquad (1.164)$$

In Equation 1.163, without indicating the limits for now, substitute the expressions

$$z = (y^2 + a^2 + h^2)^{1/2}; \quad y^2 = \{z^2 - (a^2 + h^2)\};$$

$$y = \{z^2 - (a^2 + h^2)\}^{1/2}; \quad dy = [z/\{z^2 - (a^2 + h^2)\}^{1/2}] dz$$

to obtain

$$I = (2a^2 + 3h^2) \int [\{z^2 - (a^2 + h^2)\}^{1/2} \, z/\{z^2 - (a^2 + h^2)\}^{1/2} \, \{h^2 + z^2 - (a^2 + h^2)\}^2 \, z^3\} dz \qquad (1.165)$$

Simplify to

$$I = (2a^2 + 3h^2) \int \{1/z^2 \, (z^2 - a^2)^2\} \, dz \qquad (1.166)$$

In Equation 1.166 substitute

$$z = a \sec \alpha; \quad z^2 = a^2 \sec^2 \alpha; \quad dz = a \sec \alpha \tan \alpha \, d\alpha$$

to obtain

$$I = (2a^2 + 3h^2) \int \{(a \sec \alpha \tan \alpha)/(a^6 \sec^2 \alpha \tan^4 \alpha)\} \, d\alpha \qquad (1.167)$$

Simplify to

$$I = \{(2a^2 + 3h^2)/a^5\} \int \{\cos^4 \alpha/\sin^3 \alpha\} \, d\alpha \qquad (1.168)$$

When simplified further, Equation 1.168 reduces to

$$I = \{(2a^2 + 3h^2)/a^5\} \left\{ \int \csc^3 \alpha - 2 \int \csc \alpha + \int \sin \alpha \right\} d\alpha \qquad (1.169)$$

Using standard integration formulas, Equation 1.169 is integrated as

$$I = -\{(2a^2 + 3h^2)/a^5\}[(1/2)\csc \alpha \cot \alpha + (3/2) \, Ln \, (\csc \alpha - \cot \alpha) + \cos \alpha] \qquad (1.170)$$

Next we integrate Equation 1.164 using previous substitutions without regard to the limits until the integration is complete. Equation 1.164 is first reduced to

$$II = 3 \int [\{z^2 - (a^2 + h^2)\}^{3/2}/\{z^2 - (a^2 + h^2)\}^{1/2} \, \{h^2 + z^2 - (a^2 + h^2)\}^2 \, z^3] \, z \, dz \qquad (1.171)$$

Equation 1.171 is further simplified to

$$II = 3 \left[\int \{1/(z^2 - a^2)^2\} dz - (a^2 + h^2) \int \{1/z^2(z^2 - a^2)^2\} dz \right] \qquad (1.172)$$

Integrate the terms in Equation 1.172 separately. The first term is integrated as follows:

$$3 \int \{1/(z^2 - a^2)^2\} dz = 3 \int \{(a \sec \alpha \tan \alpha \, d\alpha)/a^4 \tan^4 \alpha$$

$$= (3/a^3) \int \{\cos^2 \alpha/\sin^3 \alpha\} d\alpha$$

$$= (3/a^3) \int \{(1 - \sin^2 \alpha)/\sin^3 \alpha\} \, d\alpha$$

$$3 \int \{1/(z^2 - a^2)^2\} dz = (3/a^3) \left\{ \int \csc^3 \alpha \, d\alpha - \int \csc \alpha \, d\alpha \right\} \qquad (1.173)$$

Equation 1.173 is integrated using standard integration formulas to

$$3 \int \{1/(z^2 - a^2)^2\}dz = (3/a^3) \{-(1/2) \csc \alpha \cot \alpha - (1/2) Ln (\csc \alpha - \cot \alpha)\}$$

(1.174)

The second term is integrated by using standard integration formulas as shown below:

$$- 3(a^2 + h^2) \int \{1/z^2 (z^2 - a^2)^2\}dz = -\{3(a^2 + h^2)/a^5\} \left\{ \int \csc^3\alpha - \right.$$

$$2 \int \csc\alpha + \int \sin\alpha \right\} d\alpha = -\{3(a^2 + h^2)/a^5\}\{-(1/2) \csc \alpha \cot \alpha$$

$$- (3/2) Ln (\csc \alpha - \cot \alpha) - \cos \alpha\}$$

(1.175)

Collect terms for II as follows:

$$- (3/2) \csc \alpha \cot \alpha \{(1/a^3) - (a^2 + h^2)/a^5\}$$

$$= (3h^2/2a^5)(\csc \alpha \cot \alpha) - (3/2) Ln (\csc \alpha - \cot \alpha) \{(1/a^3) - 3(a^2 + h^2)/a^5\}$$

$$= (3/a^5) (a^2 + 3h^2) Ln (\csc \alpha - \cot \alpha)$$

$$II = (3/2a^5)\{(2a^2 + 3h^2) Ln (\csc \alpha - \cot \alpha)$$

$$+ (3/a^5)(a^2 + h^2) \cos \alpha + (3h^2/2a^5)(\csc \alpha \cot \alpha)$$

(1.176)

Combine I and II by collecting similar terms as follows:

$$(1/2) \csc \alpha \cot \alpha \{(1/a^5)(2a^2 + 3h^2) - (3h^2/a^5)\} = - (1/a^3) \csc \alpha \cot \alpha$$

$$\cos \alpha \{(-1/a^5)(2a^2 + 3h^2) + (3/a^5)(a^2 + h^2)\} = (1/a^3) \cos \alpha$$

Sum up the above terms to obtain

$$P_o = (1/a^3)(qah^3/2\pi b)\{\cos \alpha - \csc \alpha \cot \alpha\}$$

(1.177)

Transform Equation 1.177 to the variable z and write

$$P_o = (qh^3/2\pi a^2 b)\{- a^3/z (z^2 - a^2)\}$$

(1.178)

Then transform Equation 1.178 to the y variable as follows:

$$P_o = - (qh^3/2\pi a^2 b)\{a^3/(y^2 + a^2 + h^2)^{1/2} (y^2 + h^2)\}$$

(1.179)

In Equation 1.179, P_o is evaluated from the limit 0 to b as follows:

$$P_o = (qah^3/2\pi b)\{(1/h^2A) - (1/B^2C)\}$$

(1.180)

The vertical stress under the q load is obtained from the relation

$$P_q = P - P_o \qquad (1.181)$$

in which $P =$ the vertical stress under the corner of a rectangular area loaded with a uniform load q in Equation 1.139.

The total vertical stress is the integration of Equation 1.181 through depth h along the Z-axis.

Let $h = z$ and $dh = dz$.

$$\text{III} = (qa/2\pi b) \int \{z/(z^2 + a^2)^{1/2}\}dz \qquad (1.182)$$

$$\text{III} = (qa/2\pi b)(z^2 + a^2)^{1/2} \qquad (1.183)$$

Evaluate the limits of Equation 1.183 from 0 to h to obtain

$$\text{III} = (qa/2\pi b)(A - a) \qquad (1.184)$$

$$\text{IV} = - (qa/2\pi b) \int \{h^3/(h^2 + b^2)(h^2 + a^2 + b^2)^{1/2}\}dh \qquad (1.185)$$

Let $h = x$ and $dh = dx$ in Equation 1.185 and write

$$\text{IV} = - (qa/2\pi b) \int \{x^3/(x^2 + b^2)(x^2 + a^2 + b^2)^{1/2}\}dx \qquad (1.186)$$

Let

$$z^2 = x^2 + a^2 + b^2; \quad z = (x^2 + a^2 + b^2)^{1/2}; \quad x^2 = z^2 - (a^2 + b^2)^{1/2}; \quad z^3 = (a^2 + b^2 + x^2)^{3/2};$$

$$x = (z^2 - a^2 - b^2)^{1/2}; \quad x^3 = (z^2 - a^2 - b^2)^{3/2}; \quad dx = (zdz)/(z^2 - a^2 - b^2)^{1/2}$$

Substitute these expressions in Equation 1.186 to obtain

$$\text{IV} = - (qa/2\pi b) \int \{z - [b^2/(z^2 - a^2)]\}dz \qquad (1.187)$$

In the second term of Equation 1.187, let $z = a \sec \alpha$; $dz = a \sec \alpha \tan \alpha \, d\alpha$; $z^2 = a^2 \sec^2 \alpha$. Then the expression is reduced to

$$(b^2/a) \int \csc \alpha \, d\alpha = (b^2/a)\{Ln(\csc \alpha - \cot \alpha)\} = (b^2/a)\{Ln \, (z^2 - a^2)^{1/2}\}$$

Hence,

$$IV = - (qa/2\pi b) \{z - (b^2/a)Ln\ (z^2 - a^2)^{1/2}\} \qquad (1.188)$$

Transform Equation 1.188 to the x variable to obtain

$$IV = - (qa/2\pi b)\{(x^2 + a^2 + b^2)^{1/2} - (b^2/a)Ln\{[(x^2 + a^2 + b^2)^{1/2} - a]/(b^2 + x^2)\}\} \qquad (1.189)$$

Evaluate Equation 1.189 with limits from 0 to h to obtain

$$IV = - (qa/2\pi b)\{-(b^2/a)[Ln\{b(C - a)/B(D - a)\}] + (C - D)\} \qquad (1.190)$$

Add III and IV to obtain

$$P_{oT} = (qa/2\pi b)\{(A - a) + (b^2/a)Ln\ [b(C - a)/B(D - a)] - (C - D)\} \qquad (1.191)$$

Equation 1.191 is the total vertical stress under the 0 load corner. It follows that

$$P_{qT} = P_T - P_{oT} \qquad (1.192)$$

in which P_T is given by Equation 1.157. Substitute Equation 1.157 and 1.191 in Equation 1.192 to obtain

$$P_{qT} = (q/2\pi h)\{(\pi/2)h + b\ Ln\ [(D + a)(C - a)/(D - a)(C + a)]$$

$$+ a\ Ln[(D + b)\ (C - b)/(D - b)(C + b)] - h\ arctan\ (hC/ab)\}$$

$$- (qa/2\pi b)\{(A - a) + (b^2/a)Ln\ [b(C - a)/B(D - a)] - (C - D)\} \qquad (1.193)$$

1.4.2.2 Application

The derived formulas for vertical stresses caused by a triangular surface loading are applicable in calculating the magnitude of soil settlements due to surface loads as well as the magnitude of surface loads transmitted to underground structures such as tunnels, bridges, and culverts. The formulas for triangular loading are coupled with the formulas for the uniform surface load in conjunction with the principle of superposition to obtain total vertical stresses at corners and at any other point of a rectangular area under these surface loads. From these values an accurate prediction of the magnitude of soil settlement and transmitted vertical stress on the roofs of underground structures can be calculated.

Example 1.11

Determine the values and plots of the vertical stresses under the zero and q line of a rectangular area with a triangular surface loading, given the parameters a = 10 ft., b = 20 ft. and q = 1 ksf.

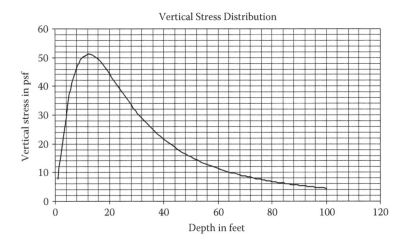

Figure 1.39 Vertical stress distribution.

Solution: Using Table 1.9 in Excel with Figures 1.39 and 1.40 and Table 1.10 with Figures 1.41 and 1.42 is the required solution to the problem.

Note that Table 1.9 with Figure 1.39 and 1.40 shows the vertical stress values and their plots under the zero load corner. Table 1.10 and Figure 1.41 and 1.42 are the vertical stress values and their plots under the q load corner.

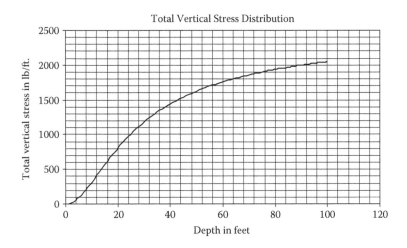

Figure 1.40 Total vertical stress distribution.

Table 1.9 Tabulated Values of Vertical Stresses

a =	10										b(C - a)/	bb/a *		
b =	20	D =	22.36											
q =	1													
h	A	B	C	E	F	G	h in ft.	Po in psf	A - a	C - D	B(D - a)	Ln(L6)	h in ft.	POT in psf
1	10.05	20.02	22.38	0.08	0.09950	0.00011	1	8	0.04988	0.02235	1.0006	0.0223	1	4
2	10.20	20.10	22.45	0.64	0.02451	0.00011	2	16	0.19804	0.08926	1.0022	0.0888	2	16
3	10.44	20.22	22.56	2.15	0.01064	0.00011	3	23	0.44031	0.20035	1.0050	0.1981	3	35
4	10.77	20.40	22.72	5.09	0.00580	0.00011	4	29	0.77033	0.35495	1.0087	0.3481	4	61
5	11.18	20.62	22.91	9.95	0.00358	0.00010	5	35	1.18034	0.55220	1.0135	0.5357	5	93
6	11.66	20.88	23.15	17.19	0.00238	0.00010	6	39	1.66190	0.79009	1.0191	0.7576	6	130
7	12.21	21.19	23.43	27.30	0.00167	0.00010	7	43	2.20656	1.07007	1.0256	1.0099	7	171
8	12.81	21.54	23.75	40.74	0.00122	0.00009	8	46	2.80625	1.38800	1.0327	1.2885	8	215
9	13.45	21.93	24.10	58.01	0.00092	0.00009	9	48	3.45362	1.74326	1.0405	1.5893	9	263
10	14.14	22.36	24.49	79.58	0.00071	0.00008	10	50	4.14214	2.13422	1.0489	1.9082	10	312
11	14.87	22.83	24.92	105.92	0.00056	0.00008	11	51	4.86607	2.55919	1.0576	2.2412	11	362
12	15.62	23.32	25.38	137.51	0.00044	0.00007	12	51	5.62050	3.01648	1.0668	2.5848	12	413
13	16.40	23.85	25.87	174.83	0.00036	0.00007	13	51	6.40122	3.50435	1.0761	2.9356	13	464
14	17.20	24.41	26.38	218.36	0.00030	0.00006	14	51	7.20465	4.02113	1.0857	3.2905	14	515
15	18.03	25.00	26.93	268.57	0.00025	0.00006	15	50	8.02776	4.56514	1.0955	3.6471	15	566
16	18.87	25.61	27.50	325.95	0.00021	0.00006	16	49	8.86796	5.13477	1.1053	4.0029	16	616
17	19.72	26.25	28.09	390.96	0.00018	0.00005	17	48	9.72308	5.72846	1.1151	4.3561	17	665
18	20.59	26.91	28.71	464.09	0.00015	0.00005	18	47	10.59126	6.34472	1.1248	4.7051	18	712
19	21.47	27.59	29.34	545.82	0.00013	0.00004	19	46	11.47091	6.98212	1.1345	5.0486	19	759
20	22.36	28.28	30.00	636.62	0.00011	0.00004	20	45	12.36068	7.63932	1.1441	5.3855	20	804
21	23.26	29.00	30.68	736.97	0.00010	0.00004	21	43	13.25941	8.31504	1.1536	5.7150	21	848
22	24.17	29.73	31.37	847.34	0.00009	0.00004	22	42	14.16609	9.00809	1.1629	6.0366	22	891
23	25.08	30.48	32.08	968.22	0.00008	0.00003	23	40	15.07987	9.71735	1.1720	6.3496	23	932
24	26.00	31.24	32.80	1100.08	0.00007	0.00003	24	39	16.00000	10.44176	1.1810	6.6539	24	972
25	26.93	32.02	33.54	1243.40	0.00006	0.00003	25	38	16.92582	11.18034	1.1897	6.9493	25	1010
26	27.86	32.80	34.29	1398.65	0.00005	0.00003	26	36	17.85678	11.93218	1.1983	7.2357	26	1047
27	28.79	33.60	35.06	1566.32	0.00005	0.00003	27	35	18.79236	12.69642	1.2066	7.5130	27	1083
28	29.73	34.41	35.83	1746.88	0.00004	0.00002	28	34	19.73214	13.47227	1.2147	7.7814	28	1117
29	30.68	35.23	36.62	1940.81	0.00004	0.00002	29	33	20.67572	14.25899	1.2227	8.0410	29	1151
30	31.62	36.06	37.42	2148.59	0.00004	0.00002	30	31	21.62278	15.05589	1.2304	8.2920	30	1182
31	32.57	36.89	38.22	2370.69	0.00003	0.00002	31	30	22.57299	15.86235	1.2378	8.5345	31	1213
32	33.53	37.74	39.04	2607.59	0.00003	0.00002	32	29	23.52611	16.67776	1.2451	8.7688	32	1243
33	34.48	38.59	39.86	2859.77	0.00003	0.00002	33	28	24.48188	17.50158	1.2522	8.9951	33	1271
34	35.44	39.45	40.69	3127.71	0.00002	0.00002	34	27	25.44009	18.33330	1.2590	9.2137	34	1299
35	36.40	40.31	41.53	3411.88	0.00002	0.00001	35	26	26.40055	19.17244	1.2657	9.4247	35	1325
36	37.36	41.18	42.38	3712.76	0.00002	0.00001	36	25	27.36308	20.01856	1.2722	9.6286	36	1351
37	38.33	42.06	43.23	4030.83	0.00002	0.00001	37	24	28.32754	20.87125	1.2784	9.8255	37	1375
38	39.29	42.94	44.09	4366.56	0.00002	0.00001	38	23	29.29377	21.73014	1.2845	10.0157	38	1399
39	40.26	43.83	44.96	4720.44	0.00002	0.00001	39	22	30.26164	22.59485	1.2904	10.1995	39	1422
40	41.23	44.72	45.83	5092.95	0.00002	0.00001	40	22	31.23106	23.46508	1.2962	10.3771	40	1444
41	42.20	45.62	46.70	5484.55	0.00001	0.00001	41	21	32.20190	24.34050	1.3018	10.5488	41	1465
42	43.17	46.52	47.58	5895.72	0.00001	0.00001	42	20	33.17407	25.22083	1.3072	10.7147	42	1486
43	44.15	47.42	48.47	6326.95	0.00001	0.00001	43	19	34.14748	26.10580	1.3124	10.8751	43	1505
44	45.12	48.33	49.36	6778.71	0.00001	0.00001	44	19	35.12206	26.99517	1.3175	11.0304	44	1524
45	46.10	49.24	50.25	7251.48	0.00001	0.00001	45	18	36.09772	27.88870	1.3225	11.1805	45	1543
46	47.07	50.16	51.15	7745.73	0.00001	0.00001	46	18	37.07441	28.78617	1.3273	11.3259	46	1561
47	48.05	51.08	52.05	8261.95	0.00001	0.00001	47	17	38.05206	29.68737	1.3320	11.4666	47	1578
48	49.03	52.00	52.95	8800.61	0.00001	0.00001	48	16	39.03060	30.59213	1.3365	11.6028	48	1595
49	50.01	52.92	53.86	9362.19	0.00001	0.00001	49	16	40.01000	31.50025	1.3409	11.7348	49	1611
50	50.99	53.85	54.77	9947.16	0.00001	0.00001	50	15	40.99020	32.41158	1.3452	11.8627	50	1627
51	51.97	54.78	55.69	10556.01	0.00001	0.00001	51	15	41.97115	33.32594	1.3494	11.9867	51	1642
52	52.95	55.71	56.60	11189.20	0.00001	0.00001	52	14	42.95281	34.24321	1.3535	12.1069	52	1657
53	53.94	56.65	57.52	11847.23	0.00001	0.00001	53	14	43.93515	35.16323	1.3574	12.2235	53	1671
54	54.92	57.58	58.45	12530.56	0.00001	0.00001	54	14	44.91812	36.08588	1.3613	12.3366	54	1685
55	55.90	58.52	59.37	13239.67	0.00001	0.00000	55	13	45.90170	37.01103	1.3650	12.4464	55	1698
56	56.89	59.46	60.30	13975.04	0.00001	0.00000	56	13	46.88585	37.93857	1.3687	12.5531	56	1711
57	57.87	60.41	61.23	14737.16	0.00001	0.00000	57	12	47.87055	38.86840	1.3722	12.6566	57	1724
58	58.86	61.35	62.16	15526.48	0.00001	0.00000	58	12	48.85576	39.80040	1.3757	12.7572	58	1736
59	59.84	62.30	63.10	16343.50	0.00000	0.00000	59	12	49.84146	40.73449	1.3790	12.8549	59	1748
60	60.83	63.25	64.03	17188.69	0.00000	0.00000	60	11	50.82763	41.67056	1.3823	12.9500	60	1759
61	61.81	64.20	64.97	18062.53	0.00000	0.00000	61	11	51.81424	42.60854	1.3855	13.0424	61	1770
62	62.80	65.15	65.91	18965.50	0.00000	0.00000	62	11	52.80127	43.54835	1.3886	13.1323	62	1781
63	63.79	66.10	66.85	19898.06	0.00000	0.00000	63	10	53.78871	44.48990	1.3917	13.2197	63	1792
64	64.78	67.05	67.79	20860.71	0.00000	0.00000	64	10	54.77654	45.43313	1.3946	13.3049	64	1802
65	65.76	68.01	68.74	21853.91	0.00000	0.00000	65	10	55.76473	46.37796	1.3975	13.3877	65	1812
66	66.75	68.96	69.69	22878.15	0.00000	0.00000	66	10	56.75328	47.32433	1.4003	13.4685	66	1822
67	67.74	69.92	70.63	23933.90	0.00000	0.00000	67	9	57.74216	48.27217	1.4031	13.5471	67	1832
68	68.73	70.88	71.58	25021.65	0.00000	0.00000	68	9	58.73136	49.22144	1.4058	13.6237	68	1841
69	69.72	71.84	72.53	26141.85	0.00000	0.00000	69	9	59.72087	50.17207	1.4084	13.6984	69	1850

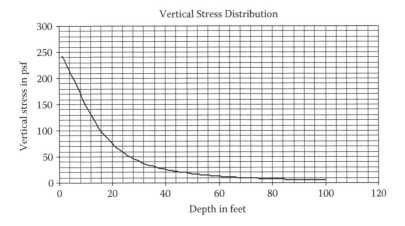

Figure 1.41 Vertical stress distribution.

1.4.3 Trapezoidal load on a rectangular area

For more than half a century, the integration of Boussinesq's elastic equation for a trapezoidal surface loading has eluded investigators on this subject. As a result, graphical and finite-element procedures are currently being utilized to solve this problem. The following derivations show the integration of the elastic equation for total vertical stress due to a trapezoidal loading by using a superposition method and the derived equations for the rectangular and triangular surface loading.

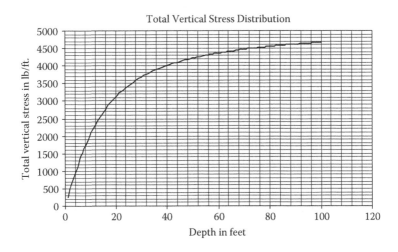

Figure 1.42 Total vertical stress distribution.

Table 1.10 Tabulated Values of Vertical Stresses

	Vertical Stress Values			Total Vertical Stress Values			
P	Po	h in ft.	Pq in psf	P_T	P_{OT}	h in ft.	P_{QT} in lb/ft.
250	8	1	242	250	4	1	246
249	16	2	234	500	16	2	484
247	23	3	225	748	35	3	713
244	29	4	215	994	61	4	933
239	35	5	205	1235	93	5	1143
233	39	6	194	1471	130	6	1342
226	43	7	183	1701	171	7	1530
218	46	8	172	1922	215	8	1707
209	48	9	161	2136	263	9	1873
200	50	10	150	2340	312	10	2029
191	51	11	140	2536	362	11	2174
182	51	12	131	2722	413	12	2309
173	51	13	122	2899	464	13	2435
164	51	14	114	3068	515	14	2553
156	50	15	106	3228	566	15	2662
148	49	16	99	3380	616	16	2765
141	48	17	92	3525	665	17	2860
133	47	18	86	3662	712	18	2949
127	46	19	81	3792	759	19	3033
120	45	20	76	3915	804	20	3111
114	43	21	71	4032	848	21	3184
108	42	22	66	4143	891	22	3253
103	40	23	62	4249	932	23	3317
98	39	24	59	4349	972	24	3378
93	38	25	55	4445	1010	25	3435
89	36	26	52	4536	1047	26	3489
84	35	27	49	4622	1083	27	3539
80	34	28	47	4705	1117	28	3587
77	33	29	44	4783	1151	29	3633
73	31	30	42	4858	1182	30	3676
70	30	31	40	4930	1213	31	3717
67	29	32	38	4998	1243	32	3755
64	28	33	36	5063	1271	33	3792
61	27	34	34	5126	1299	34	3827
59	26	35	33	5186	1325	35	3861
56	25	36	31	5243	1351	36	3892
54	24	37	30	5298	1375	37	3923
52	23	38	28	5351	1399	38	3952
49	22	39	27	5401	1422	39	3979
48	22	40	26	5450	1444	40	4006
46	21	41	25	5496	1465	41	4031
44	20	42	24	5541	1486	42	4055
42	19	43	23	5584	1505	43	4079
41	19	44	22	5626	1524	44	4101
39	18	45	21	5666	1543	45	4123
38	18	46	20	5704	1561	46	4143
36	17	47	19	5741	1578	47	4163
35	16	48	19	5777	1595	48	4182
34	16	49	18	5811	1611	49	4200
33	15	50	17	5845	1627	50	4218
32	15	51	17	5877	1642	51	4235
31	14	52	16	5908	1657	52	4252
30	14	53	16	5938	1671	53	4268
29	14	54	15	5968	1685	54	4283
28	13	55	15	5996	1698	55	4298
27	13	56	14	6023	1711	56	4312
26	12	57	14	6050	1724	57	4326
25	12	58	13	6075	1736	58	4340
25	12	59	13	6100	1748	59	4353
24	11	60	12	6124	1759	60	4365
23	11	61	12	6148	1770	61	4377
22	11	62	12	6171	1781	62	4389
22	10	63	11	6193	1792	63	4401
21	10	64	11	6214	1802	64	4412
21	10	65	11	6235	1812	65	4423
20	10	66	10	6255	1822	66	4433
19	9	67	10	6275	1832	67	4443
19	9	68	10	6294	1841	68	4453
18	9	69	10	6313	1850	69	4463
18	9	70	9	6331	1859	70	4472

1.4.3.1 Derivation

Figure 1.43 shows a trapezoidal loading common to roadway embankments. Using the principle of superposition, the vertical stress under the center of the embankment at a point h is derived as follows:

Due to MON = Equation 1.193, in which q is replaced by $bq/(b-c)$ (1.194)

Due to RQN = Equation 1.193, in which q is replaced by $cq/(b-c)$ (1.195)

Hence,

$$P_{zT} = 2(\text{MON} - \text{RQN}) = \{\text{Equation 1.194} - \text{Equation 1.195}\} \quad (1.196)$$

Similarly, the total vertical stress under point M can be derived by superposition of three triangular loads $(q_1 - q_2 - q)$ using Equation 1.191, in which

$$q_1 = 2bq/(b-c) \quad \text{with} \quad b_1 = 2b, \quad q_2 = \{q(b+c)/(b-c)\} \quad \text{with}$$
$$b_2 = b + c \quad \text{and} \quad q \quad \text{with} \quad b = b - c$$

1.4.3.2 Application

The derived formulas can determine the exact average stress valve to use in the standard settlement formula due to a surface load of a roadway embankment. Moreover, the magnitude of the transmitted surface load on the roof of any underground structure can be precisely calculated. The principle

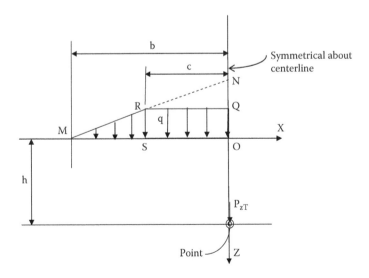

Figure 1.43 Vertical stress under a trapezoidal loading intensity.

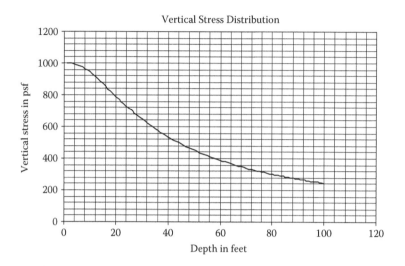

Figure 1.44 Vertical stress distribution.

of superposition must be employed in the solution for the dispersion of this trapezoidal surface load. From these values the total vertical stress on this rectangular area can be calculated to any desired degree of accuracy, and hence, an accurate prediction on the soil settlement and accurate estimate of the transmitted vertical stress on the roof of underground structures can be determined.

Example 1.12

In Figure 1.43 the embankment is 600 ft. long and has the values b = 50 ft., c = 30 ft., and q = 1 ksf. Determine the vertical stress values under the center of the embankment.

> **Solution:** Using the Excel program for uniform and triangular loading in Table 1.11 and Figures 1.44 and 1.45 is the required solution to this problem. The superpositioning of triangular loadings are as follows:

For triangle MNO the parameters are $a = 300$ ft., $b = 50$ ft., and $q = 2.50$ ksf. For triangle RQN the parameters are $a = 300$ ft., $b = 30$ ft., and $q = 1.50$ ksf. The vertical stresses under the center are obtained from 4(MNO – RQN) loadings.

Hence, two sets of Excel programs for triangular loading have to be set up before the final tabulation and plots of vertical stress distribution can be obtained.

In Figure 1.46, at any point between A and B, the vertical pressure at a depth under point G is determined from two sets of triangular loading; that is,

$$(OEG - AEF) + (CDG - BDF) \tag{1.197}$$

Table 1.11 Tabulated Values of Vertical Stresses

MNO - RQN				Total Vertical Stress Values			
Vertical Stress Values							
MNO	RQN	h in ft.	Pq in psf	MNO	RQN	h in ft.	PqT in lb/ft.
617	367	1	1000	621	371	1	1000
609	359	2	1000	1234	734	2	2000
601	351	3	1000	1839	1089	3	3000
593	343	4	1000	2436	1437	4	4000
585	336	5	999	3026	1776	5	4999
577	328	6	998	3607	2108	6	5998
570	320	7	998	4181	2432	7	6996
562	313	8	996	4746	2748	8	7993
554	305	9	995	5304	3057	9	8988
546	298	10	993	5855	3359	10	9982
539	291	11	991	6397	3654	11	10974
531	284	12	988	6932	3941	12	11964
524	277	13	986	7460	4222	13	12951
516	271	14	982	7980	4496	14	13935
509	264	15	979	8493	4764	15	14916
502	258	16	975	8998	5025	16	15893
495	252	17	971	9496	5280	17	16866
488	246	18	966	9987	5529	18	17834
481	240	19	961	10471	5772	19	18798
474	235	20	956	10948	6009	20	19756
467	229	21	950	11419	6241	21	20709
460	224	22	945	11882	6468	22	21657
453	219	23	938	12339	6689	23	22598
447	214	24	932	12789	6906	24	23534
441	209	25	925	13233	7117	25	24462
434	205	26	919	13670	7324	26	25384
428	200	27	912	14101	7526	27	26300
422	196	28	905	14526	7724	28	27208
416	192	29	897	14945	7918	29	28109
410	187	30	890	15358	8107	30	29002
404	184	31	882	15765	8293	31	29888
398	180	32	875	16166	8474	32	30767
393	176	33	867	16562	8652	33	31638
387	173	34	859	16952	8827	34	32501
382	169	35	851	17336	8997	35	33356
377	166	36	843	17716	9165	36	34203
371	163	37	835	18090	9329	37	35042
366	160	38	827	18459	9490	38	35874
361	157	39	819	18823	9648	39	36697
356	154	40	811	19181	9803	40	37513
352	151	41	804	19536	9956	41	38320
347	148	42	796	19885	10105	42	39120
342	145	43	788	20229	10252	43	39911
338	143	44	780	20570	10396	44	40695
333	140	45	772	20905	10537	45	41471
329	138	46	764	21236	10676	46	42239
325	136	47	756	21563	10813	47	42999
321	133	48	749	21886	10948	48	43752
316	131	49	741	22204	11080	49	44497
312	129	50	734	22519	11210	50	45234
308	127	51	726	22829	11338	51	45964
305	125	52	719	23135	11464	52	46687
301	123	53	711	23438	11588	53	47402
297	121	54	704	23737	11710	54	48110
293	119	55	697	24032	11830	55	48810
290	117	56	690	24324	11948	56	49504
286	116	57	683	24612	12064	57	50190
283	114	58	676	24897	12179	58	50870
280	112	59	669	25178	12292	59	51543
276	111	60	662	25456	12404	60	52208
273	109	61	656	25730	12513	61	52868
270	108	62	649	26002	12622	62	53520
267	106	63	643	26270	12729	63	54166
264	105	64	636	26535	12834	64	54806
261	103	65	630	26797	12938	65	55439
258	102	66	624	27057	13040	66	56066
255	100	67	618	27313	13141	67	56687
252	99	68	612	27566	13241	68	57301
249	98	69	606	27817	13339	69	57910
247	97	70	600	28065	13437	70	58513

in which q in each triangle is determined from the relationships such that $q_1 = qx/(b-c)$ in triangle OEG; $q_2 = q[x - (b-c)]/(b-c)$ in triangle AEF; $q_3 = q(2b-x)/(b-c)$ in triangle CDG; $q_4 = q[(b+c) - x]/(b-c)$ in triangle BDF. Hence,

$$q = (q_1 - q_2) + (q_3 - q_4) \qquad (1.198)$$

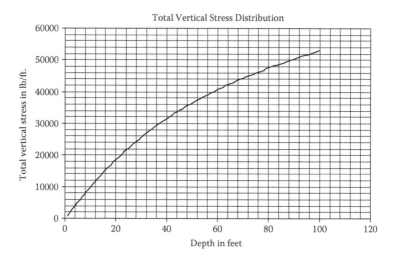

Figure 1.45 Total vertical stress distribution.

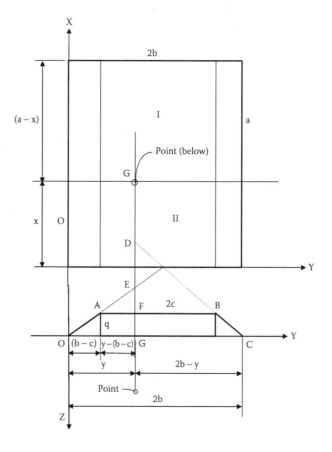

Figure 1.46 Point (between A and B) inside the loaded area.

Table 1.12 Breakdown of Areas for Superposition

Triangle	Area I Dimensions		Area II Dimensions	
	a	b	a	b
OEG	(a - x)	y	x	y
AEF	(a - x)	y - (b - c)	x	y - (b - c)
CDG	(a - x)	2b - y	x	2b - y
BDF	(a - x)	2c - [y - (b - c)]	x	2c - [y - (b - c)]

The point under consideration will divide the plan of the loaded area into two parts designated as I and II, whose area is $2b(a - x)$ and $2bx$, respectively. Substitute Equation 1.197 into Equation 1.198 to obtain the total pressure under point G for areas I and II.

Example 1.13

Using the dimensions in Example 1.12, calculate the vertical stresses under the point in Figure 1.46 when x = 200 ft., y = 30 ft., and q = 1 ksf.

> **Solution:** First construct Tables 1.12 and 1.13 for area I and area II. Then use the Excel program set up in previous example to obtain Table 1.14 and Figures 1.47 and 1.48 as the required solution to the problem.

Table 1.13 Assigned Values for Computation

Triangle	Area I			Area II		
	a	b	q	a	b	q
OEG	400	30	1.5	200	30	1.5
AEF	400	10	0.5	200	10	0.5
CDG	400	70	3.5	200	70	3.5
BDF	400	50	2.5	200	50	2.5

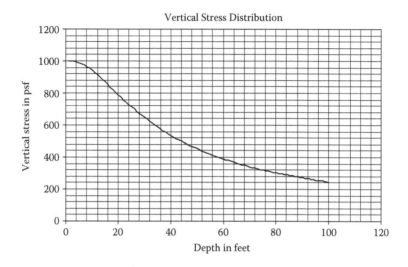

Figure 1.47 Vertical stress distribution.

In Figure 1.49, at any point between O and F, the vertical stress under point G involves the relationships of the triangles CDG, ABD, AEF, and EGO such that

$$CDG - (ABD + AEF) + EGO \qquad (1.199)$$

in which q for each triangle is determined as follows: $q_1 = q(2b - y)/(b - c)$ in triangle CDG; $q_2 = q(b + c - y)/(b - c)$ in triangle ABD; $q_3 = qy/(b - c)$ in triangle EGO; $q_4 = q(b - c - y)/(b - c)$ in triangle AEF.

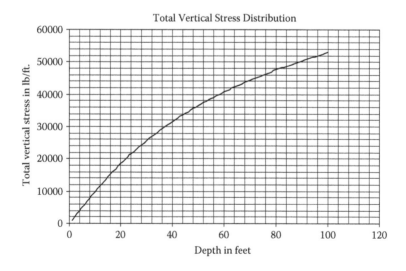

Figure 1.48 Total vertical stress distribution.

Table 1.14 Breakdown of Areas for Superposition

		TOTAL 1						GRAND TOTAL			
OEG	AEF	h in ft.	P_{ql} in psf	OEG	AEF	h in ft.	P_{qT} in lb/ft.	h in ft.	P_q in psf	h in ft.	P_{qT} in lb/ft.
734	234	1	500	742	242	1	500	1	1000	1	1000
718	219	2	500	1468	468	2	1000	2	999	2	2000
702	204	3	499	2179	679	3	1499	3	998	3	2998
687	189	4	497	2873	876	4	1997	4	995	4	3994
671	176	5	495	3552	1059	5	2493	5	990	5	4987
656	164	6	492	4215	1229	6	2987	6	984	6	5974
641	153	7	488	4864	1387	7	3477	7	976	7	6953
626	143	8	483	5497	1535	8	3962	8	966	8	7924
611	133	9	477	6115	1672	9	4442	9	955	9	8885
596	125	10	471	6718	1802	10	4917	10	943	10	9834
582	117	11	465	7308	1923	11	5385	11	930	11	10770
568	111	12	458	7883	2037	12	5846	12	915	12	11692
555	104	13	450	8444	2144	13	6300	13	901	13	12601
542	99	14	443	8992	2246	14	6747	14	886	14	13494
529	94	15	435	9528	2342	15	7186	15	870	15	14372
516	89	16	427	10050	2433	16	7617	16	854	16	15234
504	85	17	419	10560	2520	17	8040	17	838	17	16080
492	81	18	411	11058	2602	18	8455	18	823	18	16911
480	77	19	403	11544	2681	19	8863	19	807	19	17725
469	74	20	395	12019	2757	20	9262	20	791	20	18524
458	71	21	388	12482	2829	21	9654	21	775	21	19307
448	68	22	380	12935	2898	22	10037	22	760	22	20075
438	65	23	372	13378	2965	23	10414	23	745	23	20827
428	63	24	365	13811	3029	24	10782	24	730	24	21564
418	61	25	358	14234	3090	25	11144	25	715	25	22287
409	58	26	351	14648	3150	26	11498	26	701	26	22995
400	56	27	344	15052	3207	27	11845	27	687	27	23690
391	55	28	337	15448	3263	28	12185	28	674	28	24370
383	53	29	330	15835	3317	29	12518	29	660	29	25037
375	51	30	324	16214	3369	30	12845	30	648	30	25691
367	50	31	317	16585	3419	31	13166	31	635	31	26332
360	48	32	311	16948	3468	32	13480	32	623	32	26961
352	47	33	305	17304	3515	33	13789	33	611	33	27578
345	46	34	300	17653	3562	34	14091	34	599	34	28183
338	44	35	294	17995	3606	35	14388	35	588	35	28776
332	43	36	289	18329	3650	36	14679	36	577	36	29359
325	42	37	283	18658	3693	37	14965	37	566	37	29930
319	41	38	278	18980	3734	38	15246	38	556	38	30492
313	40	39	273	19296	3775	39	15521	39	546	39	31043
307	39	40	268	19606	3814	40	15792	40	536	40	31584
301	38	41	263	19910	3853	41	16058	41	527	41	32115
296	37	42	259	20209	3890	42	16319	42	518	42	32638
291	36	43	254	20502	3927	43	16575	43	509	43	33151
286	36	44	250	20791	3963	44	16828	44	500	44	33655
281	35	45	246	21074	3998	45	17076	45	492	45	34151
276	34	46	242	21352	4033	46	17319	46	483	46	34639
271	33	47	238	21625	4066	47	17559	47	475	47	35118
267	33	48	234	21894	4099	48	17795	48	468	48	35590
262	32	49	230	22158	4132	49	18027	49	460	49	36054
258	31	50	226	22418	4163	50	18255	50	453	50	36510
254	31	51	223	22674	4194	51	18480	51	446	51	36959
250	30	52	219	22926	4225	52	18701	52	439	52	37402
246	30	53	216	23173	4255	53	18919	53	432	53	37837
242	29	54	213	23417	4284	54	19133	54	426	54	38266
238	29	55	210	23657	4313	55	19344	55	419	55	38688
235	28	56	206	23893	4341	56	19552	56	413	56	39104
231	28	57	203	24126	4369	57	19757	57	407	57	39514
228	27	58	200	24356	4397	58	19959	58	401	58	39918
224	27	59	198	24581	4424	59	20158	59	395	59	40316
221	26	60	195	24804	4450	60	20354	60	390	60	40708
218	26	61	192	25024	4476	61	20548	61	384	61	41095
215	25	62	189	25240	4502	62	20738	62	379	62	41477
212	25	63	187	25453	4527	63	20926	63	374	63	41853
209	25	64	184	25664	4552	64	21112	64	369	64	42224
206	24	65	182	25871	4576	65	21295	65	364	65	42590
203	24	66	179	26076	4600	66	21476	66	359	66	42951
201	24	67	177	26278	4624	67	21654	67	354	67	43308
198	23	68	175	26477	4647	68	21830	68	349	68	43659
195	23	69	172	26673	4670	69	22003	69	345	69	44006
193	23	70	170	26868	4693	70	22175	70	341	70	44349
190	22	71	168	27059	4715	71	22344	71	336	71	44688

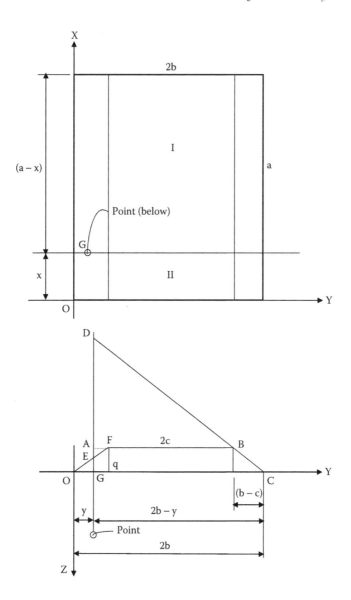

Figure 1.49 Point (between O and A) inside the loaded area.

Example 1.14

In Figure 1.49 determine the vertical stresses under the point where x = 100 ft., y = 10 ft., and other dimensions are the same as Example 1.13.

> **Solution:** Construct Table 1.14 and use the same Excel set up as in the previous example. Table 1.15 and Figures 1.50 and 1.51 are the required solutions.

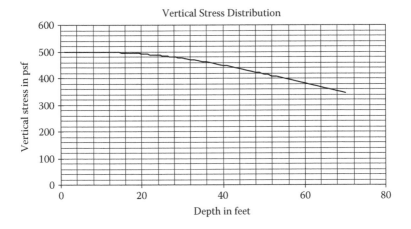

Figure 1.50 Vertical stress distribution.

In Figure 1.52, the vertical stress at a point outside the loaded area is determined by methods outlined above except that the vertical stress is obtained using superposition such that areas represented by

$$BOD + EGO - (ABC + AFG) \qquad (1.200)$$

will determine the value of the vertical stress under point O.

The value of q for each triangular loading is obtained as follows: $q_1 = (2b + y)$ $q/(b - c)$ in triangle BOD; $q_2 = (b + c + y)q/(b - c)$ in triangle ABC; $q_3 = (y + b -$

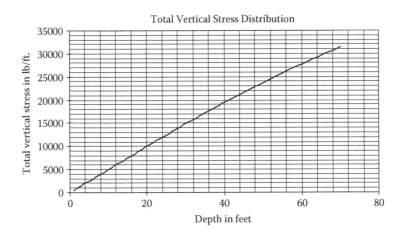

Figure 1.51 Total vertical stress distribution.

Table 1.15 Tabulated Values of Vertical Stresses

		TOTAL 1						GRAND TOTAL			
CDG	ABD	h in ft.	Pql in psf	CDG	ABD	h in ft.	PqT in lb/ft.	h in ft.	Pq in psf	h in ft.	PqT in lb/ft.
2234	1734	1	500	2242	1742	1	500	1	500	1	500
2218	1718	2	500	4468	3468	2	1000	2	500	2	1000
2202	1702	3	500	6678	5178	3	1500	3	500	3	1500
2186	1686	4	500	8873	6873	4	2000	4	500	4	2000
2170	1671	5	500	11051	8551	5	2500	5	500	5	2500
2154	1655	6	500	13213	10214	6	3000	6	500	6	3000
2139	1639	7	500	15360	11861	7	3499	7	500	7	3499
2123	1623	8	499	17491	13492	8	3999	8	499	8	3999
2107	1607	9	499	19605	15107	9	4498	9	499	9	4498
2091	1592	10	499	21704	16707	10	4997	10	499	10	4997
2075	1576	11	498	23786	18291	11	5496	11	498	11	5496
2059	1561	12	498	25853	19859	12	5994	12	498	12	5994
2043	1545	13	498	27904	21412	13	6492	13	498	13	6492
2027	1530	14	497	29938	22949	14	6989	14	497	14	6989
2011	1514	15	496	31957	24471	15	7486	15	496	15	7486
1995	1499	16	495	33960	25978	16	7981	16	495	16	7981
1979	1484	17	495	35946	27470	17	8476	17	495	17	8476
1963	1469	18	494	37917	28946	18	8971	18	494	18	8971
1946	1454	19	493	39871	30408	19	9464	19	493	19	9464
1930	1439	20	491	41810	31854	20	9956	20	491	20	9956
1914	1424	21	490	43732	33285	21	10447	21	490	21	10447
1898	1409	22	489	45638	34702	22	10936	22	489	22	10936
1882	1395	23	487	47529	36104	23	11424	23	487	23	11424
1866	1380	24	486	49403	37492	24	11911	24	486	24	11911
1850	1366	25	484	51261	38865	25	12396	25	484	25	12396
1834	1352	26	482	53103	40224	26	12879	26	482	26	12879
1818	1338	27	481	54930	41569	27	13361	27	481	27	13361
1802	1324	28	479	56740	42899	28	13841	28	479	28	13841
1786	1310	29	477	58534	44216	29	14318	29	477	29	14318
1770	1296	30	475	60313	45519	30	14794	30	475	30	14794
1755	1282	31	472	62075	46808	31	15267	31	472	31	15267
1739	1269	32	470	63822	48083	32	15739	32	470	32	15739
1723	1255	33	468	65552	49345	33	16207	33	468	33	16207
1707	1242	34	465	67267	50594	34	16674	34	465	34	16674
1691	1229	35	463	68966	51829	35	17138	35	463	35	17138
1676	1216	36	460	70650	53051	36	17599	36	460	36	17599
1660	1203	37	457	72318	54260	37	18057	37	457	37	18057
1645	1190	38	454	73970	55457	38	18513	38	454	38	18513
1629	1178	39	451	75607	56641	39	18966	39	451	39	18966
1614	1165	40	449	77228	57812	40	19416	40	449	40	19416
1598	1153	41	446	78834	58971	41	19863	41	446	41	19863
1583	1140	42	443	80425	60117	42	20307	42	443	42	20307
1568	1128	43	439	82000	61252	43	20748	43	439	43	20748
1553	1116	44	436	83560	62374	44	21186	44	436	44	21186
1538	1105	45	433	85105	63485	45	21621	45	433	45	21621
1523	1093	46	430	86635	64583	46	22052	46	430	46	22052
1508	1081	47	426	88151	65671	47	22480	47	426	47	22480
1493	1070	48	423	89651	66746	48	22905	48	423	48	22905
1478	1059	49	420	91137	67811	49	23326	49	420	49	23326
1464	1048	50	416	92608	68864	50	23744	50	416	50	23744
1449	1037	51	413	94065	69906	51	24159	51	413	51	24159
1435	1026	52	409	95507	70937	52	24570	52	409	52	24570
1421	1015	53	406	96935	71957	53	24978	53	406	53	24978
1407	1004	54	403	98349	72967	54	25382	54	403	54	25382
1393	994	55	399	99749	73966	55	25783	55	399	55	25783
1379	983	56	396	101135	74954	56	26180	56	396	56	26180
1365	973	57	392	102507	75933	57	26574	57	392	57	26574
1352	963	58	388	103865	76901	58	26964	58	388	58	26964
1338	953	59	385	105210	77859	59	27351	59	385	59	27351
1325	943	60	381	106541	78807	60	27734	60	381	60	27734
1311	934	61	378	107859	79745	61	28114	61	378	61	28114
1298	924	62	374	109164	80674	62	28490	62	374	62	28490
1285	914	63	371	110456	81593	63	28863	63	371	63	28863
1272	905	64	367	111735	82503	64	29232	64	367	64	29232
1260	896	65	364	113001	83403	65	29598	65	364	65	29598
1247	887	66	360	114254	84294	66	29960	66	360	66	29960
1235	878	67	357	115495	85177	67	30318	67	357	67	30318
1222	869	68	353	116723	86050	68	30673	68	353	68	30673
1210	860	69	350	117939	86914	69	31025	69	350	69	31025
1198	851	70	347	119143	87770	70	31373	70	347	70	31373

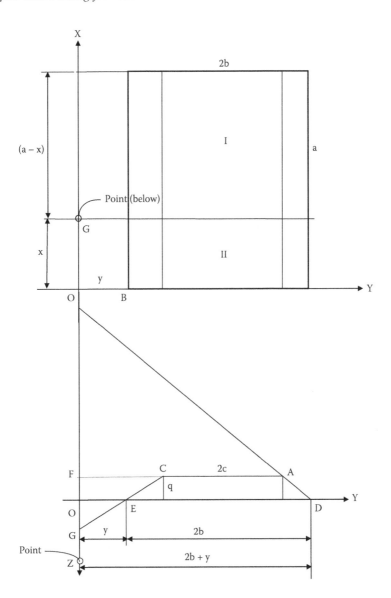

Figure 1.52 Point outside the loaded area.

$c)q/(b - c)$ in triangle AFG; $q_4 = qy/(b - c)$ in triangle EGO. Table 1.16 shows the breakdown of areas for computing the vertical stress.

Figure 1.53 shows the diagram for calculating the vertical stress at a point outside adjacent sides of the loaded area. For this case the breakdown of areas is shown in Table 1.17.

Table 1.16 Breakdown of Areas for Superposition

Triangle	q	Area I Dimensions		Area II Dimensions	
		a	b	a	b
BOD	q(2b + y)/(b - c)	(a - x)	(2b + y)	x	(2b + y)
ABC	q(b + c + y)/(b - c)	(a - x)	(b + c + y)	x	(b + c + y)
AFG	q(y + b - c)/(b - c)	(a - x)	(y + b - c)	x	(y + b - c)
EGO	qy/(b - c)	(a - x)	y	x	y

1.4.4   Uniform load on a circular area

Figure 1.54 shows the dispersion of a concentrated load Q through soil medium by using Boussinesq's elastic equation for a point load. The vertical stress is given by

$$y = (3Q/2)\{h^3/(x^2 + h^2)^{5/2}$$ (1.201)

The area of the pressure diagram can be integrated from 0 to r by using the expression

$$A = (3Q/2)h^3 \int \{dx/(x^2 + h^2)^{5/2}\}$$ (1.202)

Table 1.17 Breakdown of Areas for Superposition

Triangle	q	Area I Dimensions		Area II Dimensions	
		a	b	a	b
BOD	q(2b + y)/(b - c)	(a + x)	(2b + y)	x	(2b + y)
ABC	q(b + c + y)/(b - c)	(a + x)	(b + c + y)	x	(b + c + y)
AFG	q(y + b - c)/(b - c)	(a + x)	(y + b - c)	x	(y + b - c)
EGO	qy/(b - c)	(a + x)	y	x	y

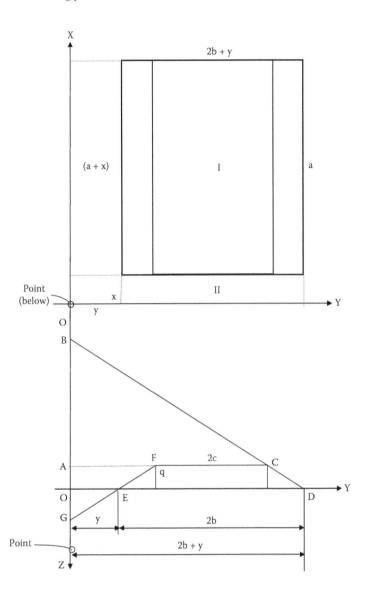

Figure 1.53 Point outside adjacent sides.

Let $x = h \tan z \, dz$ and $dx = h \sec^2 z \, dz$ and substitute these expressions in Equation 1.202 to get

$$\{dx/(x^2 + h^2)^{5/2}\} = h \sec^2 z \, dz/[h(1 + \tan^2 z)^{1/2}]^5 \tag{1.203}$$

Expand the right side of Equation 1.203 to obtain

$$\{dx/(x^2 + h^2)^{5/2}\} = (1/h^4)\{1 - \sin^2 z\}\cos z \, dz \tag{1.204}$$

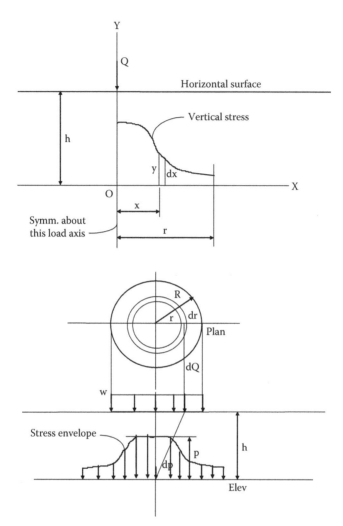

Figure 1.54 Vertical stress at the center of a circular footing.

Integrate Equation 1.201 from 0 to r and obtain

$$A = (3Q/2h)\{r/(r^2 + h^2)^{1/2}\}[1 - r^2/3(r^2 + h^2)] \tag{1.205}$$

Simplify Equation 1.205 to obtain

$$A = (Qr/2h)\{(2r^2 + 3h^2)/(r^2 + h^2)^{3/2}\} \tag{1.206}$$

The centroid of this area with respect to the load axis given by the expression

$$A\bar{x} = (3Q/2)h^3 \int \{(xdx/)x^2 + h^2)^{5/2}\} \tag{1.207}$$

Integrate Equation 1.207 from 0 to r to obtain

$$A\bar{x} = (Q/2)\{[(r^2 + h^2)^{3/2} - h^3]/(r^2 + h^2)^{3/2}\} \tag{1.208}$$

Solve for the centroid by substituting Equation 1.206 into Equation 1.208 to obtain

$$\bar{x} = [h(r^2 + h^2)^{3/2} - h^4]/r(2r^2 + 3h^2) \tag{1.209}$$

The total vertical stress under a circular footing with radius r at any horizontal plane h distance below the footing is equal to the volume of the vertical stress distribution on this circular area; that is, by Pappus' theorem,

$$V = 2\,A\bar{x} \tag{1.210}$$

Substitute Equation 1.206 into Equation 1.210 to obtain

$$V = Q\{1 - h^3/(r^2 + h^2)^{3/2}\} \tag{1.211}$$

In Equation 1.211, the concentrated load diminishes in value as we increase the depth h under the footing.

For a uniform loading w use Boussinesq's elastic equation for a point load $dQ = (2\pi\,rdr)w$ and

$$dp = 3(2\pi\,rdr)w\,h^3/2\pi\,(r^2 + h^2)^{3/2} \tag{1.212}$$

Integrate Equation 1.212 and evaluate using the limits 0 to R to obtain

$$p = w\{1 - h^3/(R^2 + h^2)^{3/2}\} \tag{1.213}$$

Equation 1.213 is the vertical stress at the center of a circular area. To solve for the total vertical stress at this point, integrate Equation 1.213 from 0 to h. First let $h = z$.

$$P_T = w \int \{1 - z^3/(z^2 + R^2)^{3/2}\}dz \tag{1.214}$$

In the second term of Equation 1.214, let $z = R \tan \alpha$; $dz = R \sec^2 \alpha\, d\alpha$; and $z^2 = R^2 \tan^2 \alpha$.

Substitute these expressions in Equation 1.211 and integrate this derivative to

$$P_T = w\{z - (z^2 + 2R^2)/(z^2 + R^2)^{1/2}\} \tag{1.215}$$

Evaluate Equation 1.215 using the limits 0 to h.

$$P_T = [w/(h^2 + R^2)^{1/2} \{(h + 2R)(h^2 + R^2)^{1/2} - 2R^2 - h^2\} + R^2) \tag{1.216}$$

$$\Delta p = [w/h(h^2 + R^2)^{1/2} \{(h + 2R)(h^2 + R^2)^{1/2} - 2R^2 - h^2\} + R^2) \tag{1.217}$$

Equations 1.216 and 1.217 are used in settlement calculations under the center of a circular footing foundation. At this stage the reader should be able to try a numerical example to determine the magnitude of Δp using these two equations. For a single layer directly underneath a circular footing, Equation 1.217 is all that is needed to solve for Δp. However, when the compressible soil layer of thickness H is overlain by other soil layers, apply Equation 1.216 at the bottom of this layer and again at the top of this layer. Get the difference and divide it by H to obtain the value of Δp to use in the standard settlement formula.

Notations

For footing design

>b: width of a rectangular footing
>c: compressive depth of a footing
>d: length of a rectangular footing
>q: allowable soil bearing pressure
>f'_c: ultimate strength of concrete
>f_y: yield strength of steel
>M: moment uplift capacity
>M_R: resultant internal moment uplift capacity
>M_u: external resultant bending moment
>M_z: component of M_R perpendicular to the capacity axis
>R: radius of a circular column
>V: vertical uplift capacity
>V_u: external vertical load
>M_o: moment due to overburden
>θ: position of the footing capacity axis with respect to the horizontal
>β: position of the resultant with respect to the capacity axis
>θ_u: position of the resultant with respect to the horizontal

Note: Refer to Jarquio 2004 other symbols and alphabets.

For settlement calculations

>Δp: average pressure at the center of a soil layer underneath a footing foundation
>σ_z: vertical stress at any point under a concentrated load, Q
>P: vertical stress at a depth h under a rectangular load
>P_T: total vertical stress at a depth h under a rectangular area loaded with q
>q: loading intensity
>P_o: vertical stress at a depth h due to a triangular load q

P_{oT}: total vertical stress at a depth h under a rectangular area loaded with a triangular load q

P_z: the vertical stress at a depth h under a trapezoidal load

P_{zT}: the total vertical stress at a depth h under a rectangular area loaded with a trapezoidal load q

Note: All other alphabets and symbols used in the mathematical derivations are defined in the context of their use in the analysis.

chapter two

Steel sections

2.1 Introduction

The analytical method described in this book precludes the use of the standard flexure and interaction formula in predicting the yield capacity of steel tubular sections. The types of tubular sections considered are steel pipes and square and rectangular tubing. The analytical method is based on the application of basic mathematics and the strength of materials approach. The stress–strain plot of the steel material is known to be linear up to its yield stress. From this property we can write the equations of the forces that can be developed on a given steel tube section using calculus. The author has demonstrated this procedure in papers presented at the ISEC-01, ISEC-02, and SEMC2001 international conferences and ASCE 2004 Structures Congress as well as in his book *Analytical Method in Reinforced Concrete* (2004). The same approach will be utilized in this book.

The steel stress diagram to use in the analysis is shown in Figure 2.1. The reference diagram is the balanced condition in which the tensile and compressive steel yield stress is developed simultaneously. This balanced condition is attained when the axial capacity of the steel section is zero and the maximum bending moment capacity is developed. As the value of the compressive depth c is increased from $h/2$, the compressive capacity of the steel section is developed. When the value of the compressive depth is less than $h/2$, the tensile capacity of the steel section is developed. Tensile stresses are above the horizontal line, while compressive stresses are below the horizontal line in Figure 2.1.

These stress diagrams are considered to act on the steel section, forming stress volumes that when calculated will yield the axial and bending moment capacity of the steel section. To determine the capacity of a steel pipe, we shall use the principle of superposition of outer and inner circular sections.

Variables considered are the radius of the steel pipe, the thickness of the shell, and the compressive depth and yield stress of the steel material. Yield capacity is measured by sets of axial and moment capacity of the tubular

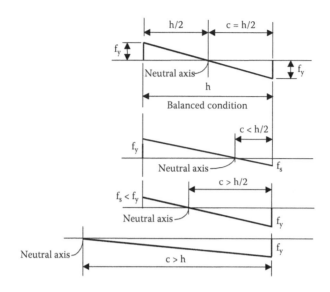

Figure 2.1 Steel stress diagrams.

steel section. The equations are programmed in a Microsoft Excel worksheet to plot the yield capacity curve of a tubular steel section.

Capacity curves for steel pipes whose dimensions are listed by the *Federal Steel and AISC Steel Manual* are constructed for easy reference by a civil or structural engineer involved in the design of steel pipes.

When these capacity curves or tables are available, the only thing left for the civil or structural engineer is the determination of the external loads. Once the external loads are known or decided, they are either plotted on the capacity curves or numerical values are compared and the most efficient section is selected to support these loads.

2.2   Steel pipe

In Figure 2.2 the variables considered are the radius of the circular tube R, the thickness of the shell t, the compressive depth c, and the yield stress f_y of the steel material. The equations of the forces and moments around the centerline of the outer circular area are derived first, followed by the inner circular area of the steel tubular section. There are three cases for the envelope of values for the compressive depth c. The balanced condition is at the center of the circular section due to symmetrical development of the compressive and tensile yield stresses of the steel material. The stress volumes on the circular section are represented as V_a and V_b, which determine the values of the forces C and T. The corresponding moments around the centerline are designated M_a and M_b, respectively. From equilibrium conditions of $\Sigma F = 0$ and $\Sigma M = 0$, the axial and moment capacities of the steel tubular section are easily obtained. Using basic mathematics, the following expressions for forces and moments are derived:

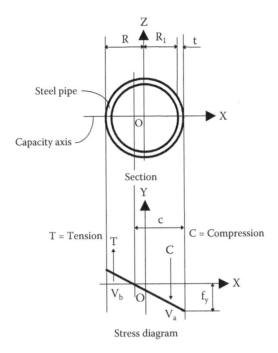

Figure 2.2 Steel pipe column section.

2.2.1   Outer circle

The equation of the circle is given by the expression $z^2 = x^2 + R^2$, from which

$$z = (R^2 - x^2)^{1/2} \tag{2.1}$$

The equation of the stress function when $c > R$ is given by the expression

$$y = (f_y/c)\{x + (c - R)\} \tag{2.2}$$

and when $c < R$, it is given by the expression

$$y = [f_y/(2R - c)]\{x + (c - R)\} \tag{2.3}$$

Case 1: $0 < c < R$

$$dV_a = 2[f_y/(2R - c)]\{R^2 - x^2\}^{1/2}\{x + (c - R)\}dx \tag{2.4}$$

Use the integration formulas as follows:

$$\{R^2 - x^2\}^{1/2}\,dx = (1/2)\{x[R^2 - x^2]^{1/2} + R^2 \arcsin (x/R)\} \tag{2.5}$$

$$\{R^2 - x^2\}^{1/2}xdx = -(1/3)\{[R^2 - x^2]^3\}^{1/2} \tag{2.6}$$

$$\{R^2 - x^2\}^{1/2}x^2dx = -(x/4)\{[R^2 - x^2]^3\}^{1/2} + (R^2/8)\{x[R^2 - x^2]^{1/2} + R^2\arcsin(x/R)\}$$
$$(2.7)$$

$$V_a = [2f_y/(2R - c)](1/3)\{[R^2 - x^2]^3\}^{1/2} + (1/2)(c - R)\{x[R^2 - x^2]^{1/2} R^2\arcsin(x/R)$$
$$(2.8)$$

Evaluate the limits in Equation 2.8 from $(R - c)$ to R to obtain

$$V_a = [2f_y/(2R - c)]\{(1/3)[(2Rc - c^2)]^{1/2} -$$
$$(1/2)(R - c)[1.5708R^2 - (R - c)(2Rc - c^2)^{1/2} -$$
$$R^2 \arcsin\{(R - c)/R\}]\}$$
$$(2.9)$$

The derivative for the moment about the centerline is given by

$$dMa = [2f_y/(2R - c)]\{[R^2 - x^2]^{1/2} x^2 + (c - R)x[R^2 - x^2]^{1/2}\}dx \qquad (2.10)$$

Integrate and evaluate the limits of Equation 2.10 from $(R-c)$ to R to obtain

$$M_a = [2f_y/(2R - c)]\{(\pi R^4/16) - (R^4/8)\arcsin[(c - R)/R]$$
$$- (R^2/8) (R - c)(2Rc - c^2)^{1/2} - (1/12)(R - c)[(2Rc - c^2)^3]^{1/2}\} \qquad (2.11)$$

$$V_b = [2f_y/(2R - c)]\{R^2 - x^2\}^{1/2}\{x + (c - R)\}dx \qquad (2.12)$$

Integrate and evaluate the limits of Equation 2.12 from $-R$ to $(R-c)$ to obtain

$$V_b = (2f_y/[2R - c])\{-(1/3)[(2R - c^2)^3]^{1/2} +$$
$$(1/2)(c - R) [1.5708R^2 + (R - c) (2Rc - c^2)^{1/2} +$$
$$R^2 \arcsin [(R - c)/R]\}$$
$$(2.13)$$

$$dM_b = [2f_y/(2R - c)]\{R^2 - x^2\}^{1/2}\{x^2 + (c - R)x\}dx \qquad (2.14)$$

Integrate and evaluate the limits of Equation 2.14 from $-R$ to $(R-c)$ to obtain

$$M_b = (2f_y/[2R - c])\{(\pi R^4/16) - (1/12)(c - R)[(2Rc - c^2)^3]^{1/2}$$
$$+ (R^4/8)\arcsin[(c - R)/R] - (R^2/8)(c - R)(2Rc - c^2)^{1/2}\} \qquad (2.15)$$

Case 2: $R < c < 2R$. Note that the same derivative and integral expressions above are used in the following derivations. Only the limits are not the same, as, for instance, under this case the limits are from $-(c - R)$ to R for V_a and M_a. Evaluate the limits and obtain

$$V_a = (2f_y/c)\{(1/3)[(2Rc - c^2)^3]^{1/2} +$$
$$(1/2)(c - R) [1.5708R^2 + (c - R)(2Rc - c^2)^{1/2} +$$
$$R^2\arcsin[(c - R)/R]\}$$
$$(2.16)$$

$$M_a = (2f_y/c)\{(\pi R^4/16) + (R^4/8)\arcsin[(c - R)/R] +$$
$$(1/12)(c - R) [(2Rc - c^2)^3]^{1/2} +$$
$$(R^2/8)(c - R)(2Rc - c^2)^{1/2}\} \tag{2.17}$$

Evaluate the above limits to obtain

$$V_b = (2f_y/c) \{(1/4)(c - R)(\pi R^2) + (1/3)[(2Rc - c^2)^3]^{1/2} +$$
$$(1/2)(c - R) [1.5708R^2 + (c - R)(2Rc - c^2)^{1/2}$$
$$+ R^2 \arcsin[(c - R)/R]\} \tag{2.18}$$

$$M_b = (2f_y/c)\{(\pi R^4/16) - (1/12)(c - R)[(2Rc - c^2)^3]^{1/2}$$
$$- (R^4/8)\arcsin [(c - R)/R] - (R^2/8)(c - R)(2Rc - c^2)^{1/2}\} \tag{2.19}$$

Case 3: $c > 2R$. The limits for V_a and M_a are $-R$ to R. Evaluate the limits to obtain

$$V_a = (2f_y/c) (c - R)(\pi R^2) \tag{2.20}$$

$$M_a = (f_y/4c)(\pi R^4) \tag{2.21}$$

2.2.2   Inner circle

Replace R with R_1 in some of the terms in the equations for the outer circle. Note that

$$R_1 = R - t \tag{2.22}$$

Case 1: $c < t$. $V_a = 0$ and $M_a = 0$.

$$dV_b = [2f_y/(2R - c)]\{(R_1^2 - x^2)^{1/2} [x + (c - R)]\}dx \tag{2.23}$$

Integrate and evaluate the limits of Equation 2.23 from $-R_1$ to R_1 to obtain

$$V_b = (f_y/[2R - c])(c - R) \pi R_1^2 \tag{2.24}$$

Similarly,

$$dM_b = [2f_y/(2R - c)](R_1^2 - x^2)^{1/2} \{x^2 + (c - R)x\}dx \tag{2.25}$$

Integrate and evaluate the limits of Equation 2.25 from $-R_1$ to R_1 to obtain

$$M_b = (1/4)(f_y/[2R - c])\pi R_1^4 \tag{2.26}$$

Case 2: $0 < c < R$. Evaluate the limits from $(R - c)$ to R_1 to obtain

$$V_a = [2f_y/(2R-c)]\left\{(1/3)\left(\left[(R_1^2-(R-c)^2\right]^3\right)^{1/2}+\right.$$

$$(1/2)(c-R)\left[1.5708R_1^2-(R-c)\left(\left[R_1^2-(R-c)^2\right]^3\right)^{1/2}-\right.$$

$$\left.R_1^2 \arcsin[(R-c)/R_1]\right\} \tag{2.27}$$

Similarly evaluate the limits from $(R - c)$ to R_1 to get

$$M_a = [2f_y(2R-c)]\left\{\left(\pi R_1^4/16\right)-\left(R_1^4/8\right)\arcsin[(c-R)/R_1]-\right.$$

$$\left(R_1^2/8\right)(R-c)\left(\left[R_1^2-(R-c)^2\right]^3\right)^{1/2}-$$

$$\left.(1/12)(R-c)\left(\left[R_1^2-(R-c)^2\right]^3\right)^{1/2}\right\} \tag{2.28}$$

Evaluate the limits from $-R_1$ to $(R - c)$ to obtain

$$V_b = (2f_y/[2R-c])\left\{(-(1/3)\left(\left[R_1^2-(R-c)^2\right]^3\right)^{1/2}+\right.$$

$$(1/2)(c-R)\left[1.5708R_1^2+(R-c)\left[R_1^2-(R-c)^2\right]^{1/2}+\right.$$

$$\left.R_1^2 \arcsin p[R(R-c)/R_1]\right\} \tag{2.29}$$

Similarly evaluate the limits from $-R_1 - (R - c)$ to get

$$M_b = (2f_y/[2R-c])\left\{\left(\pi R_1^4/16\right)-(1/12)(c-R)\left(\left[R_1^2-(R-c)^2\right]^3\right)^{1/2}\right.$$

$$\left.+\left(R_1^2/8\right)\arcsin[(R-c)/R_1]-\left(R_1^2/8\right)(c-R)\left[R_1^2-(R-c)^2\right]^{1/2}\right\} \tag{2.30}$$

Case 3: $R < c < (R + R_1)$. Evaluate the limits from $-(c - R)$ to R_1 to obtain

$$V_a = (2f_y/c)\left\{(1/3)\left(\left[R_1^2-(c-R)^2\right]^3\right)^{1/2}+\right.$$

$$(1/2)(c-R)\left[1.5708R_1^2+(c-R)\left[R_1^2-(c-R)^2\right]^{1/2}+\right.$$

$$\left.R_1^2 \arcsin[(c-R)/R_1]\right\} \tag{2.31}$$

Similarly evaluate the limits from $-(c - R)$ to R_1 to get

$$M_a = (2f_y/c)\left\{\left(\pi R_1^4/16\right) + \left(R_1^4/8\right)\arcsin[(c - R)/R_1] + \right.$$
$$(1/12)(c - R)\left(\left[R_1^2 - (c - R)^2\right]^3\right)^{1/2} +$$
$$\left. (R_1^2/8)(c - R)\left[R_1^2 - (c - R)^2\right]^{1/2}\right\} \tag{2.32}$$

Evaluate the limits from $-R_1$ to $-(c - R)$ to obtain

$$V_b = (2f_y/c)\left\{(1/4)(c - R)\left(\pi R_1^2\right) + (1/3)\left(\left[R_1^2 - (c - R)^2\right]^3\right)^{1/2} \right.$$
$$+ (1/2)(c - R)\left[1.5708R_1^2 + (c - R)\left[R_1^2 - \right.\right.$$
$$\left.\left.(c - R)^2\right]^{1/2} + R_1^2 \arcsin[(c - R)/R_1]\right\} \tag{2.33}$$

Similarly evaluate the limits from $-R_1$ to $-(c - R)$ to get

$$M_b = (2f_y/c)\left\{\left(\pi R_1^4/16\right) - (1/12)(c - R)([R_1^2 - (c - R)^2]^3)^{1/2} \right.$$
$$\left. -\left(R_1^4/8\right)\arcsin[(c - R)/R_1] - \left(R_1^2/8\right)(c - R)\left[R_1^2 - (c - R)^2\right]^{1/2}\right\} \tag{2.34}$$

Case 4: $c > (R + R_1)$. Evaluate the limits from $-R_1$ to R_1 to obtain

$$V_a = (f_y/c)(c - R)\left(\pi R_1^2\right) \tag{2.35}$$

$$M_a = (f_y/4c)\left(\pi R_1^4\right) \tag{2.36}$$

From the preceding sets of equations, apply the summation of forces and moments to obtain the axial and moment capacities for the outer and inner circular sections. Get the difference of the outer and inner sections to obtain the axial capacity

$$P = \text{Sum } V_a - \text{Sum } V_b \tag{2.37}$$

and moment capacity

$$M = \text{Sum } M_a + \text{Sum } M_b \tag{2.38}$$

2.2.3   Capacity curves

Capacity curves are plotted from the values predicted by the preceding equations. The yield capacity curve of a standard-weight pipe with a 9.625 in. diameter is used to illustrate the application of these equations using an Excel 97 worksheet. The pipe's dimensions are taken from Federal Steel Supply, Inc. as $R = 4.8125$ in. (122.2 mm) and $t = 0.342$ in. (8.7 mm). Use $f_y = 36$ ksi (248 MPa).

From the Excel worksheet, Figure 2.3 is the plot of the yield capacity of this steel pipe section at all envelopes of its compressive depth c. The same graph is applicable to a pipe section when the predominant stress is tension. It is just a matter of labeling, since the same equations are applicable. Note that values below the zero line are indicated for conditions of compressive depth less than the balanced condition. The balanced condition is where the axial capacity is zero and moment value is a maximum. This is the case for beams subjected to bending loads only.

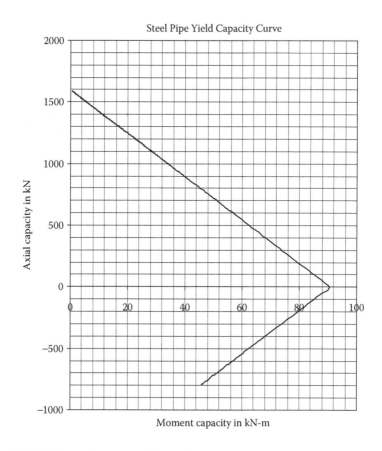

Figure 2.3 Yield capacity curve of a steel pipe.

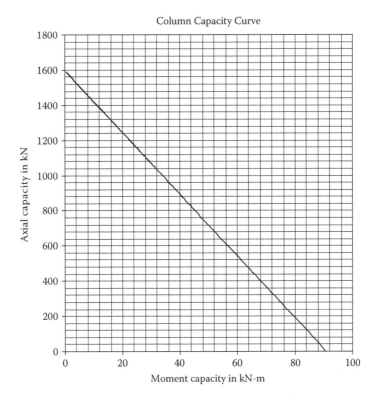

Figure 2.4 Steel pipe column capacity curve.

For a column without tension (i.e., the pipe section is in compression), the values above the zero line are compared with the external loads that a designer considers for his or her design. Figure 2.4 shows the column capacity curve for this example application of the analytical method in predicting the yield capacity of a pipe section for a column design. To use the capacity curve, all that is needed is to determine the external loads that the pipe will be subjected to. Plot the set or sets of external axial and bending moment loads on the capacity curve. Then determine the locations of the plotted values with respect to the capacity curve.

When these external loads are within the envelope of the capacity curve, the chosen steel pipe section is considered sufficient to support the external loads. The decision as to where inside the curve to limit the external loads is the responsibility of the civil or structural engineer. Every civil and structural engineer can choose all the multipliers to the external loads, such as moment magnification for slender columns, accidental eccentricities, multipliers to live loads, code requirements, safety factors, and other factors deemed applicable to a particular situation.

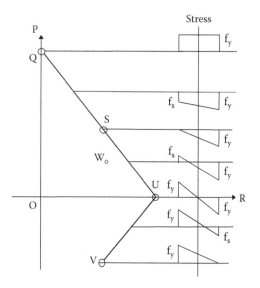

Figure 2.5 Schematic locations of key points on the capacity curve.

2.2.3.1 Key points

Figure 2.5 shows the schematic locations of the key points on the capacity curve. Two important key points are Q and U, which are the endpoints of the straight-line plot of the yield capacity of the steel pipe section. Key point U is the position of the compressive depth c at the center of the steel pipe section. It is the so-called balanced condition of loading when the compressive and tensile yield stress of the steel is developed at the same time. Key point Q is the condition when the compressive depth c approaches infinity (or a very large number). At that instant, the distribution of compressive stress becomes uniform and the value of the moment capacity approaches zero. The axial capacity is then simply the area of the section times the yield stress of steel, that is,

$$P = \pi \left(R^2 - R_1^2 \right) f_y \tag{2.39}$$

At key point U the value of the axial capacity is zero because of symmetry in the compressive and tensile stress distribution. The maximum moment capacity is developed at this point and is given by

$$M = (\pi/4) f_y \left\{ R^3 - \left(R_1^4 / R \right) \right\} \tag{2.40}$$

A straight line used to obtain the compression zone with the horizontal and vertical axis can then connect key points Q and U.

Table 2.1 Values at Key Points

Key Point	c, inches	M, in-Kips	P, Kips	c, m	M, kN-m	P, kN
Q	very large	0	359	very large	0	1597
S	9.63	402	180	24.45	45	799
U	4.81	804	0	12.22	91	0
V	0.00	402	-180	0.00	45	-798

Key point S is where the compressive depth is equal to h, the start of full compressive stress on the steel pipe section. The values of the axial and moment capacities at this point are given by the following equations:

$$P = (\pi f_y/2)\{R^2 - R_1^2\} \tag{2.41}$$

$$M = (\pi f_y/8)\{R^3 - (R_1^4/R)\} \tag{2.42}$$

Key point V is the limit of the capacity curve in the tension zone. This is attained when the compressive depth becomes very small, that is, less than t. When the value of c approaches zero, the moment capacity is given by Equation 2.42 and the axial capacity is given by Equation 2.41 with opposite sign.

Point W_o is a typical illustration of the plot of a set of the estimated external axial and bending moment that a given section is subjected to.

For columns with predominant tensile capacities, simply changing the label "Compression Zone" to "Tension Zone" can use the same plot of the column capacity curve. The values at key points for the example pipe are shown in Table 2.1.

2.2.4 Capacities of the pipe section at other stresses

Since the steel stress is a multiplier for the integral expressions in Equations 2.4 to 2.36 and can vary upward to yield stress in a linear fashion, steel pipe capacities at other stresses (including yield) can be obtained by proportion as follows:

$$M_1 = (f_s/f_y)\,M \tag{2.43}$$

$$P_1 = (f_s/f_y)P \tag{2.44}$$

From the previous example, at key point U, M is given by Equation 2.40.

$M = 0.7854\,(36)\{4.8125^3 - (4.4705^4/4.8125)\} = 805$ in-kips (91kN-m)

Then when $f_y = 50$ ksi (344.5 MPa), the moment capacity at key point U from Equation 2.43 is equal to

$M_1 = (344.5/248)(91) = 126$ kN-m

Similarly, Equations 2.32 and 2.33 give the axial and moment capacities at key point S as follows:

$$P = 1.5708(36)\{4.8125^2 - 4.4705^2\} = 180 \text{ kips (799 kN)}$$

$$M = 0.3927(36)\{4.8125^3 - (4.4705^4/4.8125)\} = 402 \text{ in-kips (45 kN-m)}$$

From Equations 2.43 and 2.44, obtain the values when $f_y = 50$ ksi (345 MPa) as follows:

$$M_1 = (344.5/248)(45) = 62 \text{ kN-m}$$

$$P_1 = (344.5/248)(799) = 1{,}110 \text{ kN}$$

For columns, key point Q is difficult if not impossible to attain due to inherent out-of-plumb from the vertical axis of any column and also because of buckling at mid-height of the column length. Because of the unavoidable eccentric loading of external loads on a column section, the minimum axial capacity may be limited at key point S.

In the preceding example column, the eccentricity of the internal capacity is given by

$$e = M/P \tag{2.45}$$

Substituting numerical values,

$$e = 402/180 = 2.23 \text{ in. (5.67 cm)}$$

The value of P at key point S can also be set as the limit of the total compressive force to be applied at bearing plates for columns. The usefulness of the key point Q when combined with key point S is the completion of the column capacity envelope (triangular area) formed with the vertical and horizontal axis of the capacity curve.

2.2.5   Reference tables

Reference tables can be constructed that indicate the capacities at key points Q, S, and U for standard weight, extra strong and double extra strong steel pipes when $f_y = 36$ ksi (248 MPa). These values are very handy for quick reference in the design of steel pipes. For values of f_y other than $f_y = 36$ ksi (248 MPa) in these tables, use Equations 2.43 and 2.44 to obtain the corresponding values of P and M.

Figure 2.6 indicates the relative positions of key points Q, S, and U in a straight line defining the capacity curve for a steel pipe. With these values as a reference, it should be easier for the structural engineer to compare the

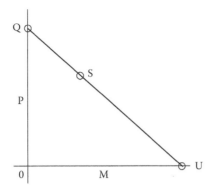

Figure 2.6 Relative positions of key points.

external loads to determine the adequacy of the size of pipe chosen to support these loads. Figure 2.7 indicates the dimensions for calculating pipe capacity.

We can also determine the actual stress, f_s, in the section by using the relationship

$$f_s = (M_u/M)f_y \tag{2.46}$$

Equation 2.45 indicates the position of the resultant internal capacity of the section. If the column is loaded such that there is no movement at the end supports, the eccentricity represents the translation of the section from the vertical axis. Due to buckling of the column, axis of column out-of-plumb, and unanticipated eccentricities of the axial load on the column, additional moments are introduced, resulting in the reduction of the axial capacity of the column. These additional eccentricities are determined as external loads since the internal capacities are functions of the geometry of the section and the given yield strength of the steel material.

Tables 2.2, 2.3, and 2.4 are for $f_y = 36$ ksi. Tables 2.5, 2.6, and 2.7 are for $f_y = 50$ ksi. From the values shown in the tabulations, we can check the direct

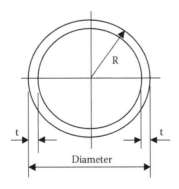

Figure 2.7 Dimensions of a steel pipe.

Table 2.2 Capacities of Standard Weight Pipe

		STANDARD WEIGHT STEEL PIPE											
		ENGLISH UNITS (M, in IN-KIPS ; P, in KIPS)						SI UNITS (M, in kN-m; P, in kN)					
fy, ksi =	36	KEY POINT Q		KEY POINT S		KEY POINT U		KEY POINT Q		KEY POINT S		KEY POINT U	
D, inches	t, inch	M	P	M	P	M	P	M	P	M	P	M	P
3.50	0.216	0	80	31	0	517	0	0	357	4	0	58	0
4.00	0.226	0	96	43	0	765	0	0	429	5	0	86	0
4.50	0.237	0	114	58	0	1082	0	0	508	7	0	122	0
5.00	0.247	0	133	75	0	1476	0	0	591	8	0	167	0
5.56	0.258	0	155	98	0	2022	0	0	689	11	0	228	0
6.63	0.280	0	201	153	0	3389	0	0	894	17	0	383	0
7.63	0.301	0	249	220	0	5140	0	0	1109	25	0	581	0
8.63	0.322	0	302	303	0	7408	0	0	1345	34	0	837	0
9.63	0.342	0	359	402	0	10259	0	0	1597	45	0	1159	0
10.75	0.365	0	429	538	0	14248	0	0	1907	61	0	1610	0
11.75	0.375	0	482	665	0	18530	0	0	2146	75	0	2093	0
12.75	0.375	0	525	789	0	23554	0	0	2335	89	0	2661	0
14.00	0.375	0	578	959	0	31011	0	0	2570	108	0	3504	0
16.00	0.375	0	663	1265	0	45959	0	0	2948	143	0	5192	0
18.00	0.375	0	748	1613	0	65063	0	0	3325	182	0	7351	0
20.00	0.375	0	832	2004	0	88832	0	0	3702	226	0	10036	0
22.00	0.375	0	917	2438	0	117775	0	0	4079	275	0	13306	0
24.00	0.375	0	1002	2913	0	152401	0	0	4457	329	0	17218	0
26.00	0.375	0	1087	3432	0	193220	0	0	4834	388	0	21830	0
28.00	0.375	0	1172	3992	0	240739	0	0	5211	451	0	27199	0
30.00	0.375	0	1256	4595	0	295469	0	0	5589	519	0	33382	0
32.00	0.375	0	1341	5241	0	357917	0	0	5966	592	0	40437	0
36.00	0.375	0	1511	6659	0	508007	0	0	6721	752	0	57394	0
42.00	0.375	0	1765	9104	0	803756	0	0	7852	1029	0	90808	0
48.00	0.375	0	2020	11931	0	1196458	0	0	8984	1348	0	135175	0

Table 2.3 Capacities of Extra Strong Pipe

		EXTRA STRONG STEEL PIPE											
		ENGLISH UNITS (M, in IN-KIPS ; P, in KIPS)						SI UNITS (M, in kN-m; P, in kN)					
fy, ksi =	36	KEY POINT Q		KEY POINT S		KEY POINT U		KEY POINT Q		KEY POINT S		KEY POINT U	
D, inches	t, inch	M	P	M	P	M	P	M	P	M	P	M	P
3.50	0.300	0	109	40	0	535	0	0	483	5	0	60	0
4.00	0.318	0	132	57	0	792	0	0	589	6	0	89	0
4.50	0.337	0	159	77	0	1120	0	0	706	9	0	127	0
5.00	0.355	0	186	101	0	1528	0	0	830	11	0	173	0
5.56	0.375	0	220	134	0	2093	0	0	979	15	0	236	0
6.63	0.432	0	303	220	0	3523	0	0	1346	25	0	398	0
7.63	0.500	0	403	337	0	5374	0	0	1792	38	0	607	0
8.63	0.500	0	459	441	0	7686	0	0	2044	50	0	868	0
9.63	0.500	0	516	560	0	10574	0	0	2295	63	0	1195	0
10.75	0.500	0	580	710	0	14592	0	0	2578	80	0	1649	0
11.75	0.500	0	636	858	0	18917	0	0	2830	97	0	2137	0
12.75	0.500	0	693	1021	0	24018	0	0	3081	115	0	2714	0
14.00	0.500	0	763	1244	0	31582	0	0	3396	141	0	3568	0
16.00	0.500	0	877	1647	0	46723	0	0	3899	186	0	5279	0
18.00	0.500	0	990	2106	0	66049	0	0	4402	238	0	7462	0
20.00	0.500	0	1103	2622	0	90068	0	0	4905	296	0	10176	0
22.00	0.500	0	1216	3195	0	119290	0	0	5408	361	0	13477	0
24.00	0.500	0	1329	3824	0	154223	0	0	5911	432	0	17424	0
26.00	0.500	0	1442	4510	0	195376	0	0	6414	510	0	22073	0
28.00	0.500	0	1555	5252	0	243259	0	0	6917	593	0	27483	0
30.00	0.500	0	1668	6051	0	298380	0	0	7420	684	0	33711	0
32.00	0.500	0	1781	6906	0	361248	0	0	7923	780	0	40813	0
36.00	0.500	0	2007	8786	0	512261	0	0	8929	993	0	57875	0
42.00	0.500	0	2347	12031	0	809609	0	0	10438	1359	0	91469	0
48.00	0.500	0	2686	15784	0	1204164	0	0	11948	1783	0	136045	0

Table 2.4 Capacities of Double Extra Strong Pipe

		DOUBLE EXTRA STRONG STEEL PIPE										
		ENGLISH UNITS (M, in IN-KIPS ; P, in KIPS)						SI UNITS (M, in kN-m; P, in kN)				
fy, ksi =	36	KEY POINT Q		KEY POINT S		KEY POINT U		KEY POINT Q		KEY POINT S		KEY POINT U
D, inches	t, inch	M	P	M	P	M	P	M	P	M	P	M	P
3.50	0.600	0	197	62	0	578	0	0	875	7	0	65	0
4.50	0.674	0	292	122	0	1211	0	0	1297	14	0	137	0
5.00	0.710	0	344	163	0	1651	0	0	1532	18	0	187	0
5.56	0.750	0	408	218	0	2261	0	0	1816	25	0	255	0
6.63	0.864	0	563	360	0	3804	0	0	2504	41	0	430	0
7.63	0.875	0	668	507	0	5715	0	0	2971	57	0	646	0
26.00	2.000	0	5429	15138	0	216632	0	0	24147	1710	0	24475	0
28.00	2.000	0	5881	17853	0	268461	0	0	26159	2017	0	30331	0
30.00	2.000	0	6333	20795	0	327868	0	0	28171	2349	0	37042	0
32.00	2.000	0	6786	23963	0	395361	0	0	30183	2707	0	44667	0
36.00	2.000	0	7691	30976	0	556641	0	0	34208	3500	0	62889	0
42.00	2.000	0	9048	43193	0	871933	0	0	40245	4880	0	98510	0
48.00	2.000	0	10405	57444	0	1287484	0	0	46281	6490	0	145459	0

proportionality of the values with respect to the changes in the values of the yield stress of steel.

In Table 2.7 at key point Q for a 3.50-in.-diameter pipe, $P = 273$ kN when $f_y = 50$ ksi, while in Table 2.4, for the same pipe with $f_y = 36$ ksi, $P = 197$ kN. Checking the proportionality, obtain the following identical ratios of the axial

Table 2.5 Capacities of Standard Weight Pipe

		STANDARD WEIGHT STEEL PIPE										
		ENGLISH UNITS (M, in IN-KIPS ; P, in KIPS)						SI UNITS (M, in kN-m; P, in kN)				
fy, ksi =	50	KEY POINT Q		KEY POINT S		KEY POINT U		KEY POINT Q		KEY POINT S		KEY POINT U
D, inches	t, inch	M	P	M	P	M	P	M	P	M	P	M	P
3.50	0.216	0	111	43	0	718	0	0	496	5	0	81	0
4.00	0.226	0	134	60	0	1062	0	0	596	7	0	120	0
4.50	0.237	0	159	80	0	1503	0	0	706	9	0	170	0
5.00	0.247	0	184	104	0	2050	0	0	820	12	0	232	0
5.56	0.258	0	215	136	0	2808	0	0	956	15	0	317	0
6.63	0.280	0	279	212	0	4707	0	0	1241	24	0	532	0
7.63	0.301	0	346	305	0	7139	0	0	1540	34	0	807	0
8.63	0.322	0	420	420	0	10289	0	0	1868	47	0	1162	0
9.63	0.342	0	499	559	0	14249	0	0	2218	63	0	1610	0
10.75	0.365	0	595	748	0	19790	0	0	2648	84	0	2236	0
11.75	0.375	0	670	923	0	25736	0	0	2980	104	0	2908	0
12.75	0.375	0	729	1095	0	32714	0	0	3242	124	0	3696	0
14.00	0.375	0	803	1331	0	43071	0	0	3570	150	0	4866	0
16.00	0.375	0	920	1757	0	63832	0	0	4094	198	0	7212	0
18.00	0.375	0	1038	2241	0	90365	0	0	4618	253	0	10209	0
20.00	0.375	0	1156	2784	0	123377	0	0	5142	314	0	13939	0
22.00	0.375	0	1274	3386	0	163576	0	0	5666	383	0	18481	0
24.00	0.375	0	1392	4046	0	211669	0	0	6190	457	0	23914	0
26.00	0.375	0	1509	4766	0	268361	0	0	6714	538	0	30319	0
28.00	0.375	0	1627	5545	0	334360	0	0	7238	626	0	37776	0
30.00	0.375	0	1745	6382	0	410374	0	0	7762	721	0	46364	0
32.00	0.375	0	1863	7279	0	497108	0	0	8286	822	0	56163	0
36.00	0.375	0	2098	9249	0	705565	0	0	9334	1045	0	79714	0
42.00	0.375	0	2452	12645	0	1116328	0	0	10906	1429	0	126122	0
48.00	0.375	0	2805	16571	0	1661748	0	0	12478	1872	0	187743	0

Table 2.6 Capacities of Extra Strong Pipe

		EXTRA STRONG STEEL PIPE										
		ENGLISH UNITS (M, in IN-KIPS ; P, in KIPS)						SI UNITS (M, in kN-m; P, in kN)				
fy, ksi =	50	KEY POINT Q		KEY POINT S		KEY POINT U		KEY POINT Q		KEY POINT S		KEY PO
D, inches	t, inch	M	P	M	P	M	P	M	P	M	P	M
3.50	0.300	0	151	56	0	743	0	0	671	6	0	84
4.00	0.318	0	184	79	0	1099	0	0	818	9	0	124
4.50	0.337	0	220	107	0	1555	0	0	980	12	0	176
5.00	0.355	0	259	141	0	2122	0	0	1152	16	0	240
5.56	0.375	0	306	186	0	2907	0	0	1359	21	0	328
6.63	0.432	0	420	306	0	4893	0	0	1869	35	0	553
7.63	0.500	0	560	468	0	7464	0	0	2489	53	0	843
8.63	0.500	0	638	613	0	10674	0	0	2838	69	0	1206
9.63	0.500	0	717	777	0	14686	0	0	3188	88	0	1659
10.75	0.500	0	805	986	0	20266	0	0	3581	111	0	2290
11.75	0.500	0	884	1192	0	26273	0	0	3930	135	0	2968
12.75	0.500	0	962	1418	0	33358	0	0	4279	160	0	3769
14.00	0.500	0	1060	1728	0	43864	0	0	4716	195	0	4956
16.00	0.500	0	1217	2287	0	64893	0	0	5415	258	0	7332
18.00	0.500	0	1374	2925	0	91734	0	0	6114	331	0	10364
20.00	0.500	0	1532	3642	0	125094	0	0	6812	411	0	14133
22.00	0.500	0	1689	4437	0	165680	0	0	7511	501	0	18718
24.00	0.500	0	1846	5311	0	214198	0	0	8210	600	0	24200
26.00	0.500	0	2003	6263	0	271356	0	0	8908	708	0	30657
28.00	0.500	0	2160	7294	0	337859	0	0	9607	824	0	38171
30.00	0.500	0	2317	8404	0	414416	0	0	10306	949	0	46820
32.00	0.500	0	2474	9592	0	501733	0	0	11004	1084	0	56685
36.00	0.500	0	2788	12203	0	711474	0	0	12402	1379	0	80382
42.00	0.500	0	3259	16709	0	1124457	0	0	14498	1888	0	127040
48.00	0.500	0	3731	21922	0	1672450	0	0	16594	2477	0	188952

capacity corresponding to the ratio of the yield stress for the example steel pipe.

$$(50/36) = (273/197) = 1.38 \quad \text{Q.E.D.}$$

With these tabulated values of steel pipe capacities as a reference, checking the adequacy of any design by comparing the external loads acting on the chosen pipe becomes easy and straightforward.

Table 2.7 Capacities of Double Extra Strong Pipe

		DOUBLE EXTRA STRONG STEEL PIPE											
		ENGLISH UNITS (M, in IN-KIPS ; P, in KIPS)						SI UNITS (M, in kN-m; P, in kN)					
fy, ksi =	50	KEY POINT Q		KEY POINT S		KEY POINT U		KEY POINT Q		KEY POINT S		KEY POINT U	
D, inches	t, inch	M	P	M	P	M	P	M	P	M	P	M	P
3.50	0.600	0	273	86	0	803	0	0	1216	10	0	91	0
4.50	0.674	0	405	170	0	1682	0	0	1802	19	0	190	0
5.00	0.710	0	478	226	0	2293	0	0	2128	26	0	259	0
5.56	0.750	0	567	302	0	3140	0	0	2522	34	0	355	0
6.63	0.864	0	782	501	0	5283	0	0	3478	57	0	597	0
7.63	0.875	0	928	705	0	7938	0	0	4127	80	0	897	0
26.00	2.000	0	7540	21025	0	300878	0	0	33537	2375	0	33993	0
28.00	2.000	0	8168	24796	0	372863	0	0	36332	2801	0	42126	0
30.00	2.000	0	8796	28882	0	455372	0	0	39127	3263	0	51448	0
32.00	2.000	0	9425	33281	0	549112	0	0	41922	3760	0	62038	0
36.00	2.000	0	10681	43022	0	773113	0	0	47511	4861	0	87346	0
42.00	2.000	0	12566	59990	0	1211018	0	0	55895	6778	0	136820	0
48.00	2.000	0	14451	79784	0	1788173	0	0	64280	9014	0	202026	0

The difficulty will be when the structural engineer must decide on the magnitude of external loads to be applied in the design. This decision will depend on the engineer's expertise and the application of building codes for the design of steel pipes as structural members.

2.3 Rectangular steel tubing

The analytical method used here is based on the application of basic mathematics and the strength of materials approach. The stress–strain plot of the steel material is known to be linear up to its yield stress. From this property we can write the equations of the forces that can be developed on a given rectangular steel tube section using calculus. The stress volumes are designated as V_1, V_2, V_3, and their corresponding components. The values of tension T and compression C are functions of the stress volumes, which are derived as described in the following.

This section describes the analytical method of predicting the yield capacity of a rectangular steel tubular section subjected to axial and biaxial bending loads. It is based on the linear stress–strain plot of the steel material when subjected to these loads. The resulting expressions for steel forces and moments produce the yield capacity curve of the rectangular steel tubular section. The analytical method precludes the use of the interaction formula for biaxial bending. Variables considered are the width b and depth d of the rectangular steel tube, the thickness of the shell, the yield stress f_y of the steel material, and the position of the capacity axis of the rectangular tubular section. The capacity axis of the tubular section has to be taken into consideration, and the optimum position of this axis for biaxial bending is chosen along the diagonal of the outer rectangular section. The equations are programmed in a Microsoft Excel worksheet to plot the yield capacity curve of a rectangular steel tubular section. An example of a yield capacity curve for a typical rectangular steel tubular section is shown to compare current methodologies in predicting the yield strength of a rectangular steel tubular section.

Figure 2.8 shows the variables to be considered to determine the yield capacity of a rectangular tubular section. The equations of the forces and moments around the centerline of the outer rectangular area are derived first, followed by the inner rectangular area of the steel tubular section. The following expressions for these forces and moments can be obtained using basic mathematics.

2.3.1 Derivation

2.3.1.1 Outer rectangular area

The equations of the lines bounding the rectangle were shown in Chapter 1 and are listed here again for deriving the equations applicable to rectangular tubing.

$$z_1 = -\tan \theta \, (x - h/2) + z_o; \quad z_2 = \cot \theta \, (x - h/2) + z_o;$$

$$z_3 = -\tan \theta \, (x + h/2) - z_o; \quad z_4 = \cot \theta \, (x + h/2) - z_o$$

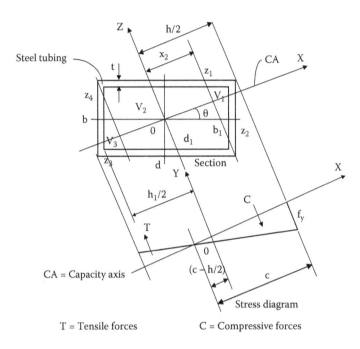

Figure 2.8 Rectangular steel tubular section.

The equation of the stress function when $c > h/2$ is given by the expression $y = (f_y/c)\{x + (c - h/2)\}$. When $c < h/2$, it is given by the expression $y = [f_y/(h - c)]\{x + (c - h/2)\}$.

Using the above expressions, we shall derive the value of the axial capacity P and then the moment capacity M.

2.3.1.2 Axial capacity derivations

Case 1: $0 < c < ((h/2) - x_2)$. See Figure 2.9 for the stress–strain diagram.

$$dV_{1a} = -(\cot\theta + \tan\theta)[f_y/(h - c)][x + (c - h/2)]dx \qquad (2.47)$$

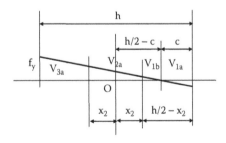

Figure 2.9 Stress–strain diagram.

Integrate and evaluate the limits of Equation 2.47 from $(h/2 - c)$ to $h/2$ to obtain

$$V_{1a} = -[f_y/6(h - c)](\cot\theta + \tan\theta)\{(h^2/4)(h - 3c) - (h/2 - c)^2(h + c)\} \quad (2.48)$$

$$dV_{1b} = -(\cot\theta + \tan\theta)[f_y/(h - c)][x + (c - h/2)]dx \quad (2.49)$$

Integrate and evaluate the limits of Equation 2.49 from x_2 to $(h/2 - c)$ to obtain

$$V_{1b} = [f_y/6(h - c)](\cot\theta + \tan\theta)\Big\{(h/2 - c)^2(c + h) -$$
$$x_2\Big[2x_2^2 + 3x_2(c - h) + 3h(c - h/2)\Big]\Big\} \quad (2.50)$$

$$dV_{2a} = (h\tan\theta + 2z_o)[f_y/(h - c)][x + (c - h/2)]dx \quad (2.51)$$

Integrate and evaluate the limits of Equation 2.51 from $-x_2$ to x_2 to get

$$V_{2a} = -[f_y/(h - c)](h\tan\theta + 2z_o)(c - h/2)2x_2 \quad (2.52)$$

$$dV_{3a} = [f_y/(h - c)](\cot\theta + \tan\theta)[x^2 + cx + (h/2)(c - h/2)]dx \quad (2.53)$$

Integrate and evaluate the limits of Equation 2.53 from $-(h/2)$ to $-x_2$ to obtain

$$V_{3a} = -[f_y/6(h - c)](\cot\theta + \tan\theta)\Big\{x_2\Big[-2x_2^2 +$$
$$3cx_2 - 3h(c - h/2)\Big] - (h^2/4)(2h - 3c)\Big\} \quad (2.54)$$

$$P = V_{1a} - V_{2a} - V_{3a} - V_{1b} \quad (2.55)$$

Case 2: $(h/2 - x_2) < c < h/2$. See Figure 2.10 for the stress–strain diagram.

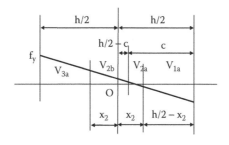

Figure 2.10 Stress–strain diagram.

V_{1a} is integrated as above but with limits from x_2 to $h/2$ to obtain

$$V_{1a} = -[f_y/6(h-c)](\cot\theta+\tan\theta)\left\{(h^2/4)(h-3c) - \right.$$
$$\left. x_2\left[2x_2^2 + 3(c-h)x_2 - 3h(c-h/2)\right]\right\} \qquad (2.56)$$

V_{2b} is integrated as above with limits from $(h/2-c)$ to x_2 to obtain

$$V_{2b} = [f_y/2(h-c)](h\tan\theta+2z_o)\{x_2[x_2+2(c-h/2)]+(h/2-c)^2\} \quad (2.57)$$

Similarly, V_{2a} is integrated with limits from $-x_2$ to $(h/2-c)$ to get

$$V_{2a} = [f_y/2(h-c)]\{(h/2-c)^2 + x_2[x_2-2(c-h/2)]\} \qquad (2.58)$$

V_{3a} = Equation 2.54.

$$P = V_{1a} + V_{2b} - V_{2a} - V_{3a} \qquad (2.59)$$

Case 3: $h/2 < c < (h/2+x_2)$. See Figure 2.11 for the stress–strain diagram.

V_{1a} = Equation 2.56 with the denominator $6c$ instead of $6(h-c)$.
V_{2a} is integrated as above with limits from $-x_2$ to $-(c-h/2)$ and the denominator changed to obtain

$$V_{2a} = (f_y/2c)(h\tan\theta+2z_o)\{(c-h/2)^2 + x_2[x_2-2(c-h/2)]\} \qquad (2.60)$$

V_{2b} = Equation 2.57 with the denominator $2c$ instead of $2(h-c)$.
V_{3a} is the same as before, except that the denominator is changed to obtain

$$V_{3a} = (f_y/6c)(\cot\theta+\tan\theta)\left\{x_2\left[-2x_2^2 + 3cx_2 - 3h(c-h/2)\right] - (h^2/4)(2h-3c)\right\}$$
$$(2.61)$$

$$P = V_{1a} + V_{2b} - V_{2a} - V_{3a} \qquad (2.62)$$

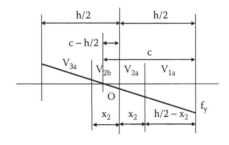

Figure 2.11 Stress–strain diagram.

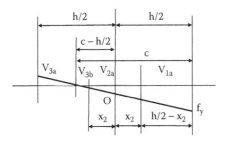

Figure 2.12 Stress–strain diagram.

Case 4: $(h/2 + x_2) < c < h$. See Figure 2.12 for the stress–strain diagram.

V_{1a} = Equation 2.56 with the denominator $6c$ instead of $6(h - c)$.
V_{2b} is integrated as above with limits from $-x_2$ to x_2 to obtain

$$V_{2b} = (f_y/c)2x_2 \, (c - h/2)(h \tan \theta + 2z_o) \tag{2.63}$$

V_{3b} is integrated as above with limits from $-(c - h/2)$ to $-x_2$ to get

$$V_{3b} = (f_y/6c)(\cot\theta + \tan\theta)\left\{x_2\left[-2x_2^2 + \right.\right.$$
$$\left.\left. 3cx_2 - 3h(c - h/2)\right] - (c - h/2)^2(c - 2h)\right\} \tag{2.64}$$

Similarly, V_{3a} is integrated with limits from $-h/2$ to $-(c - h/2)$ to get

$$V_{3a} = -(f_y/6c)(\cot\theta + \tan\theta)\{(c - h/2)^2(c - 2h) - (h^2/4)(2h - 3c)\} \tag{2.65}$$

$$P = V_{1a} + V_{2b} + V_{3b} - V_{3a} \tag{2.66}$$

Case 5: $c > h$. See Figure 2.13 for the stress–strain diagram.

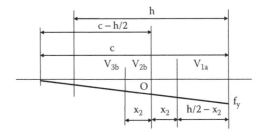

Figure 2.13 Stress–strain diagram.

V_{1a} = Equation 2.56 with the denominator $6c$ instead of $6(h - c)$. V_{2b} = Equation 2.63. V_{3b} is integrated with limits from $-h/2$ to $-x_2$ to obtain

$$V_{3b} = (f_y/6c)(\cot\theta + \tan\theta)\left\{x_2\left[-2x_2^2 + 3cx_2 - 3h(c - h/2)\right] - (h^2/4)(2h - 3c)\right\}$$

(2.67)

$$P = V_{1a} + V_{2b} + V_{3b}$$

(2.68)

The above equations will yield the axial capacity of the rectangular tubing at any position of the capacity axis from zero to α.

2.3.1.3 Moment capacity derivations
Case 1: $0 < c < (h/2 - x_2)$.

$$dV_{1a}x_{1a} = -[f_y/(h - c)](\cot\theta + \tan\theta)\{x^3 + (c - h)x^2 - (h/2)(c - h/2)x\}dx$$

(2.69)

Integrate and evaluate the limits of Equation 2.69 from $(h/2 - c)$ to $h/2$ to obtain

$$V_{1a}x_{1a} = -[f_y/12(h - c)](\cot\theta + \tan\theta)\{(h^3/16)(h - 4c) - (h/2 - c)^3(h/2 + c)\}$$

(2.70)

$$dV_{1b}x_{1b} = [f_y/(h - c)](\cot\theta + \tan\theta)\{x^3 + (c - h)x^2 - (h/2)(c - h/2)x\}dx \quad (2.71)$$

Integrate and evaluate the limits of Equation 2.71 from x_2 to $(h/2 - c)$ to obtain

$$V_{1b}x_{1b} = [f_y/12(h - c)](\cot\theta + \tan\theta)\left\{(h/2 - c)^3(h/2 + c)\right.$$
$$\left. - x_2^2\left[3x_2^2 + 4(c - h)x_2 - 3h(c - h/2)\right]\right\}$$

(2.72)

$$dV_{2a}x_{2a} = (h\tan\theta + 2z_o)[f_y/(h - c)][x^2 + (c - h/2)x]dx$$

(2.73)

Integrate and evaluate the limits of Equation 2.73 from $-x_2$ to $(h/2 - c)$ to obtain

$$V_{2a}x_{2a} = [f_y/3(h - c)](h\tan\theta + 2z + o)2x_2^3$$

(2.74)

$$dV_{3a}x_{3a} = -[fy/(h - c)](\cot\theta + \tan\theta)[x^3 + cx^2 + (h/2)(c - h/2)x]dx \quad (2.75)$$

Integrate and evaluate the limits of Equation 2.75 from $-h/2$ to $-x_2$ to obtain

$$V_{3a}x_{3a} = [f_y/12(h-c)](\cot\theta + \tan\theta)\left\{x_2^2\left[3x_2^2 - 4cx_2 + 3h(c-h/2)\right] - (h^3/4)(c-3h/4)\right\}$$

(2.76)

$$M = V_{1a}x_{1a} - V_{1b}x_{1b} + V_{2a}x2a + V_{3a}x_{3a}$$

(2.77)

Case 2: $(h/2 - x_2) < c < h/2$. $V_{1a}x_{1a}$ is integrated as above with limits from x_2 to $h/2$ to obtain

$$V_{1a}x_{1a} = -[f_y/12(h-c)](\cot\theta + \tan\theta)\left\{(h^3/16)(h-4c) - x_2^2\left[3x_2^2 + 4(c-h)x_2 - 3h(c-h/2)\right]\right\}$$

(2.78)

$V_{2b}x_{2b}$ is integrated with limits from $(h/2 - c)$ to x_2 to obtain

$$V_{2b}x_{2b} = [f_y/6(h-c)](h\tan\theta + 2z_o)\left\{x_2^2\left[2x_2 + 3(c-h/2)\right] + (h/2-c)^3\right\}$$

(2.79)

$V_{2a}x_{2a}$ is integrated with limits from $-x_2$ to $(h/2 - c)$ to get

$$V_{2a}x_{2a} = -[f_y/6(h-c)]\left\{(h/2-c)^3 + x_2^2\left[-2x_2 + 3(c-h/2)\right]\right\}$$

(2.80)

$V_{3a}x_{3a}$ = Equation 2.76.

$$M = V_{1a}x_{1a} + V_{2b}x_{2b} + V_{2a}x_{2a} + V_{3a}x_{3a}$$

(2.81)

Case 3: $h/2 < c < (h/2 + x_2)$. $V_{1a}x_{1a}$ = Equation 2.78. $V_{2a}x_{2a}$ is integrated with limits from $-x_2$ to $-(c-h/2)$ to obtain

$$V_{2a}x_{2a} = -(f_y/6c)(h\tan\theta + 2z_o)\left\{(c-h/2)^3 - x_2^2[-2x_2 + 3(c-h/2)]\right\}$$

(2.82)

$V_{3a}x_{3a}$ is the same as above except that the denominator is changed to obtain

$$V_{3a}x_{3a} = (f_y/12c)(\cot\theta + \tan\theta)\left\{x_2^2\left[3x_2^2 - 4cx_2 + 3h(c-h/2)\right] - (h^3/4)(c-3h/4)\right\}$$

(2.83)

$V_{2b}x_{2b}$ is integrated with limits from $-(c - h/2)$ to x_2 to get

$$V_{2b}x_{2b} = (f_y/6c)(h\tan\ \theta + 2z_o)\left\{x_2^2[2x_2 + 3(c-h/2)] - (c-h/2)^3\right\} \qquad (2.84)$$

$$M = V_{1a}x_{1a} + V_{2b}x_{2b} + V_{2a}x_{2a} + V_{3a}x_{3a} \qquad (2.85)$$

Case 4: $(h/2 + x_2) < c < h$. $V_{1a}x_{1a} =$ Equation 2.78. $V_{2b}x_{2b}$ is integrated with limits $-x_2$ to x_2 to obtain

$$V_{2b}x_{2b} = (f_y/3c)(h\tan\ \theta + 2z_o)\ 2x_2^3 \qquad (2.86)$$

$V_{3b}x_{3b}$ is integrated with limits from $-(c - h/2)$ to $-x_2$ to get

$$V_{3b}x_{3b} = -(f_y/12c)(\cot\ \theta + \tan\ \theta)\left\{x_2^2\left[3x_2^2 - 4cx_2 + 3h(c-h/2)\right] - (c-h/2)^3(3h/2 - c)\right\} \qquad (2.87)$$

$V_{3a}x_{3a}$ is integrated with limits from $-h/2$ to $-(c - h/2)$ to obtain

$$V_{3a}x_{3a} = (f_y/12c)(\cot\ \theta + \tan\ \theta)\{(c-h/2)^3(3h/2 - c) - (h^3/4)(c - 3h/4)\} \qquad (2.88)$$

$$M = V_{1a}x_{1a} + V_{2b}x_{2b} - V_{3b}x_{3b} + V_{3a}x_{3a} \qquad (2.89)$$

Case 5: $c > h$. $V_{1a}x_{1a} =$ Equation 2.78, and $V_{2b}x_{2b} =$ Equation 2.86. $V_{3b}x_{3b}$ is integrated with limits from $-h/2$ to $-x_2$ to get

$$V_{3b}x_{3b} = -(f_y/12c)(\cot\ \theta + \tan\ \theta)\left\{x_2^2\left[3x_2^2 - 4cx_2 + 3h(c-h/2)\right] - (h^3/4)(c - 3h/4)\right\} \qquad (2.90)$$

$$M = V_{1a}x_{1a} + V_{2b}x_{2b} - V_{3b}x_{3b} \qquad (2.91)$$

2.3.1.4   Inner rectangular area

2.3.1.4.1   Axial capacity derivations.
Case 1: $0 < c < (h/2 - h_1/2)$.

$$V_{1a} = 0 \qquad (2.92)$$

$$dV_{1b} = -[f_y/(h - c)](\cot\ \theta + \tan\ \theta)\ \{x^2 + [c - (1/2)(h + h_1)]x - (1/2)h_1(c - h/2)\}dx \qquad (2.93)$$

Integrate and evaluate the limits of Equation 2.93 from x_1 to $h_1/2$ to obtain

$$V_{1b} = [K_1/6(h-c)]\{h_1^2/4)(1.5h - 3c - 0.50h_1) -$$

$$x_1\left[2x_1^2 + 3x_1(c - 0.50h - 0.50h_1) - 3h_1(c - h/2)\right]\} \tag{2.94}$$

$$dV_{2a} = [f_y/(h-c)](h\tan\theta + 2z_o)[x + (c - h/2)]dx \tag{2.95}$$

Integrate and evaluate the limits of Equation 2.95 from $-x_1$ to x_1 to obtain

$$V_{2a} = -[K_2/(h-c)]2x_1(c - h/2) \tag{2.96}$$

$$dV_{3a} = -[f_y/(h-c)](\cot\theta + \tan\theta)\{x^2 +$$

$$[c - (1/2)(h - h_1)]x + (1/2)h_1(c - h/2)\}dx \tag{2.97}$$

Integrate and evaluate the limits of Equation 2.97 from $-h/2$ to $-x_1$ to obtain

$$V_{3a} = -[K1/6(h-c)]\{x_1\left[[-2x_1^2 + 3x_1 (c - 0.50h + 0.50h_1)\right.$$

$$\left. - 3h_1(c - h/2)\right]\left(h_1^2/4\right)(1.5h + 0.50h_1\_3c)\} \tag{2.98}$$

$$P = V_{1a} - V_{1b} - V_{2a} - V_{3a} \tag{2.99}$$

Case 2: $(h/2 - h_1/2) < c < (h/2 - x_1)$. V_{1a} is integrated with limits from $(h/2 - c)$ to $h_1/2$ to obtain

$$V_{1a} = [K_1/6(h-c)]\{(h_1^2/4)[1.5h - 3c - 0.50h_1] - (h/2 - c)^2[1.5h_1 - 0.50h + c]\} \tag{2.100}$$

Similarly, V_{1b} is integrated with limits from x_1 to $(h/2 - c)$ to get

$$V_{1b} = [K_1/6(h-c)]\{(h/2 - c)^2 [1.5h_1 + c - 0.50h]$$

$$- x_1\left[2x_1^2 + 3x_1(c - 0.50h - 0.50h_1) - 3h_1(c - h/2)\right]\} \tag{2.101}$$

V_{2a} = Equation 2.96 and V_{3a} = Equation 2.98.

$$P = V_{1a} - V_{1b} - V_{2a} - V_{3a} \tag{2.102}$$

Case 3: $(h/2 - x_1) < c < h/2$. V_{1a} is integrated with limits from $(h/2 - c)$ to $h_1/2$ to obtain

$$V_{1a} = -[K_1/6(h-c)]\left\{(h_1^2/4)(1.5h - 3c - 0.50h_1) - x_1\left[2x_1^2\right.\right.$$
$$\left.\left. + 3(c - 0.50h - 0.50h_1)x_1 - 3h_1(c - h/2)\right]\right\} \tag{2.103}$$

V_{2b} is integrated with limits from $(h/2 - c)$ to x_1 to obtain

$$V_{2b} = [K_2/2(h-c)]\{x_1[x_1 + 2(c - h/2)] + (h/2 - c)^2\} \tag{2.104}$$

Similarly, V_{2a} is integrated with limits from $-x_1$ to $(h/2 - c)$ to obtain

$$V_{2a} = [K_2/2(h-c)]\{(h/2 - c)^2 + x_1[x_1 - 2(c - h/2)]\} \tag{2.105}$$

V_{3a} = Equation 2.98.

$$P = V_{1a} + V_{2b} - V_{2a} - V_{3a} \tag{2.106}$$

Case 4: $h/2 < c < (h/2 + x_1)$. V_{1a} = Equation 2.103 and V_{2b} = Equation 2.104. V_{2a} is integrated from $-x_1$ to $-(c - h/2)$ to get

$$V_{2a} = (K_2/2c)\{(c - h/2)^2 + x_1[x_1 - 2(c - h/2)]\} \tag{2.107}$$

V_{3a} is the same as above except that the denominator has been changed to obtain

$$V_{3a} = (K_1/6c)\left\{x_1\left[-2x_1^2 + 3x_1(c - 0.50h + 0.50h_1) - 3h_1(c - h/2)\right]\right.$$
$$\left. - (h_1^2/4)(1.5h + 0.50h_1\_3c)\right\} \tag{2.108}$$

$$P = V_{1a} + V_{2b} - V_{2a} - V_{3a} \tag{2.109}$$

Case 5: $(h/2 + x_1) < c < (h_1/2 + h/2)$. V_{1a} = Equation 2.103.

$$V_{2b} = (K_2/c)2x_1 (c - h/2) \tag{2.110}$$

V_{3a} is integrated with limits from $-h_1/2$ to $-(c - h/2)$ to get

$$V_{3a} = (K_1/6c)\left\{(c - h/2)^2 [c - 0.50h - 1.5h_1] - (h_1^2/4)[0.50\,h_1 + 1.5h - 3c]\right\}$$
$$\tag{2.111}$$

Similarly, V_{3b} is integrated with limits from $-(c - h/2)$ to $-x_1$ to obtain

$$V_{3b} = (K_1/6c)\left\{x_1\left[-2x_1^2 + 3x_1\ (c - 0.50h + 0.50h_1) - 3h_1\ (c - h/2)]\right]\right.$$
$$\left. - (c - h/2)^2[c - 0.50h - 1.5h_1]\right\} \tag{2.112}$$

$$P = V_{1a} + V_{2b} - V_{3a} + V_{3b} \tag{2.113}$$

Case 4: $c > (h/2 + h_1/2)$. V_{1a} = Equation 2.103 and V_{2b} = Equation 2.110. V_{3b} is integrated with limits from $-h_1/2$ to $-x_1$ to get

$$V_{3b} = (K_1/6c)\left\{x_1\left[-2x_1^2 + 3x_1(c - 0.50h + 0.50h_1) - 3h_1\ (c - h/2)]\right]\right.$$
$$\left. - \left(h_1^2/4\right)(1.5h + 0.50h_1 - 3c)\right\} \tag{2.114}$$

$$P = V_{1a} + V_{2b} + V_{3b} \tag{2.115}$$

2.3.1.4.2 Moments capacity derivations.
Case 1: $0 < c < (h/2 - h_1/2)$.

$$V_{1a}x_{1a} = 0 \tag{2.116}$$

$$dV_{1b}x_{1b} = -[f_y/(h - c)](\cot\theta + \tan\theta)\ \{x^3 +$$
$$[c - (1/2)(h + h_1)]x^2 - (1/2)h_1(c - h/2)x\}dx \tag{2.117}$$

Integrate and evaluate the limits of Equation 2.117 from x_1 to $(h/2 - c)$ to obtain

$$V_{1b}x_{1b} = [K_1/12(h - c)]\left\{\left(h_1^3/16\right)(2h - 4c - h_1) - x_1^2\left[3x_1^2\right.\right.$$
$$\left.\left. + 4x_1(c - 0.50h - 0.50h_1) - 3h_1(c - h/2)]\right]\right\} \tag{2.118}$$

$$dV_{2a}x_{2a} = [f_y/(h - c)](h\tan\theta + 2z_o)[x^2 + (c - h/2)x]dx \tag{2.119}$$

Integrate and evaluate the limits of Equation 2.119 from $-x_1$ to x_1 to obtain

$$V_{2a}x_{2a} = [K_2/3(h - c)]2x_1^3 \tag{2.120}$$

$$dV_{3a}x_{3a} = [f_y/(h - c)](\cot\theta + \tan\theta)\ \{x^3 +$$
$$[c - (1/2)(h - h_1)]x^2 + (1/2)h_1(c - h/2)x\}dx \tag{2.121}$$

Integrate and evaluate the limits of Equation 2.121 from $-h_1/2$ to $-x_1$ to obtain

$$
V_{3a}x_{3a} = [K_1/12(h-c)]\{x_1^2[3x_1^2 - 4x_1(c-0.50h+0.50h_1)
$$
$$
+ 3h_1(c-h/2)] - (h_1^3/4)[c-0.25h_1-0.50h]\}
$$

(2.122)

$$
M = V_{3a}x_{3a} + V_{2a}x_{2a} + V_{1a}x_{1a} - V_{1b}x_{1b}
$$

(2.123)

Case 2: $(h/2 - h_1/2) < c < (h/2 - x_1)$. $V_{1a}x_{1a}$ is integrated with limits from $(h/2 - c)$ to $h_1/2$ to obtain

$$
V_{1a}x_{1a} = [K_1/12(h-c)]\left\{\left(h_1^3/4\right)(0.50h-c-0.25h_1) - (h/2-c)^3(c-0.50h+h_1)\right\}
$$

(2.124)

Similarly, $V_{1b}x_{1b}$ is integrated with limits from x_1 to $(h/2 - c)$ to obtain

$$
V_{1b}x_{1b} = [K_1/12(h-c)]\{(h/2-c)^3[c-0.50h+h_1] -
$$
$$
x_1^2\left[3x_1^2 + 4x_1(c-0.50h-0.50h_1) - 3h_1(c-h/2)\right]\}
$$

(2.125)

$V_{2a}x_{2a}$ is integrated with limits from $-x_1$ to x to obtain

$V_{2a}x_{2a} = $ Equation 2.120.

$$
V_{3a}x_{3a} = [K_1/12(h-c)]\left\{x_1^2\left[3x_1^2 - 4x_1(c-0.50h+0.50h_1)\right.\right.
$$
$$
+ 3h_1(c-h/2)\Big] - \left(h_1^3/4\right)[c-0.25h_1-0.50h]\right\}
$$

(2.126)

$$
M = V_{3a}x_{3a} + V_{2a}x_{2a} + V_{1a}x_{1a} - V_{1b}x_{1b}
$$

(2.127)

Case 3: $(h/2 - x_1) < c < h/2$. $V_{1a}x_{1a}$ is integrated with limits from x_1 to $h_1/2$ to get

$$
V_{1a}x_{1a} = [K_1/12(h-c)]\left\{\left(h_1^3/4\right)(0.50h-c-0.25h_1) -
$$
$$
x_1^2\left[3x_1^2 + 4x(c-0.50h-0.50h_1) - 3h_1(c-h/2)\right]\right\}
$$

(2.128)

$V_{2b}x_{2b}$ is integrated with limits from $(h/2 - c)$ to x_1 to get

$$
V_{2b}x_{2b} = [K_2/6(h-c)]\{x_1^2[2x_1 + 3(c-h/2)] + (h/2-c)^3\}
$$

(2.129)

$V_{2a}x_{2a}$ is integrated with limits from $-x_1$ to $(h/2 - c)$ to get

$$V_{2a}x_{2a} = -[K_2/6(h-c)]\{(h/2-c)^3 + x_1^2[2x_1 + 3(c-h/2)]\} \qquad (2.130)$$

$V_{3a}x_{3a} = $ Equation 2.126.

$$M = V_{1a}x_{1a} + V_{2b}x_{2b} + V_{2a}x_{2a} + V_{3a}x_{3a} \qquad (2.131)$$

Case 4: $h/2 < c < (h/2 + x_1)$. $V_{1a}x_{1a} = $ Equation 2.128, and $V_{3a}x_{3a}$ is the same as above except that the denominator has been changed to obtain

$$V_{3a}x_{3a} = (K_1/12c)\{x_1^2[3x_1^2 - 4x_1(c-0.50h+0.50h_1)$$
$$+ 3h_1(c-h/2)] - (h_1^3/4)(c-0.25h_1-0.50h)\} \qquad (2.132)$$

$V_{2a}x_{2a}$ is integrated with limits from $-x_1$ to $-(c - h/2)$ to get

$$V_{2a}x_{2a} = (K_2/6c)\{(c-h/2)^3 - x_1^2[-2x_1 + 3(c-h/2)]\} \qquad (2.133)$$

$V_{2b}x_{2b}$ is integrated with limits from $-(c - h/2)$ to ξ_1 to get

$$V_{2b}x_{2b} = (K_2/6c)\{x_1^2[2x_1 + 3(c-h/2)] - (c-h/2)^3\} \qquad (2.134)$$

$$M = V_{1a}x_{1a} + V_{2b}x_{2b} + V_{2a}x_{2a} + V_{3a}x_{3a} \qquad (2.135)$$

Case 5: $(h/2 + x_1) < c < (h_1/2 + h/2)$. $V_{1a}x_{1a} = $ Equation 2.128, and $V_{2b}x_{2b}$ is integrated with limits from $-x_1$ to x_1 to get

$$V_{2b}x_{2b} = (K_2/3c)2x_1^3 \qquad (2.136)$$

$V_{3b}x_{3b}$ is integrated with limits from $-(c - h/2)$ to $-x_1$ to get

$$V_{3b}x_{3b} = (K_1/12c)\{x_1^2[3x_1^2 - 4x_1(c-0.50h+0.50h_1)$$
$$+ 3h_1(c-h/2)] - (c-h/2)^3(0.50h+h_1-c)\} \qquad (2.137)$$

$V_{3a}x_{3a}$ is integrated with limits from $-h_1/2$ to $-(c - h/2)$ to get

$$V_{3a}x_{3a} = (K_1/12c)\{(c-h/2)^3(0.50h+h_1-c) - (h_1^3/4)(c-0.25h_1-0.50h)\} \qquad (2.138)$$

$$M = V_{1a}x_{1a} + V_{2b}x_{2b} - V_{3b}x_{3b} + V_{3a}x_{3a} \qquad (2.139)$$

Case 6: $c > (h/2 + h_1/2)$. $V_{1a}x_{1a}$ = Equation 2.128 and $V_{2b}x_{2b}$ = Equation 2.136. $V_{3b}x_{3b}$ is integrated with limits from $-h_1/2$ to $-x_1$ to get

$$V_{3b}x_{3b} = (K_1/12c)\left\{x_1^2\left[3x_1^2 - 4x_1(c-0.50h+0.50h_1)\right.\right.$$
$$\left.\left. + 3h_1(c-h/2)\right] - \left(h_1^3/4\right)(c-0.25h_1-0.50h)\right\}$$

(2.140)

$$M = V_{1a}x_{1a} + V_{2bx}2b - V_{3b}x_{3b}$$

(2.141)

in which

$$K_1 = (\cot\theta + \tan\theta)$$

(2.142)

$$K_2 = (h\tan\theta + 2z_0)$$

(2.143)

$$x_2 = (1/2)(d\sin\theta - b\cos\theta)$$

(2.144)

$$x_1 = (1/2)(d_1\sin\theta - b_1\cos\theta)$$

(2.145)

$$h = b\cos\theta + d\sin\theta$$

(2.146)

$$h_1 = b_1\cos\theta + d_1\sin\theta$$

(2.147)

$$b_1 = b - 2t$$

(2.148)

$$d_1 = d - 2t$$

(2.149)

$$z_0 = (1/2)(b\cos\theta - d\sin\theta)$$

(2.150)

$$z_{01} = (1/2)(b_1\cos\theta - d_1\sin\theta)$$

(2.151)

Equations 2.147 to 2.151 yield the capacity of the section along the capacity axis, measured as M (moment capacity) and P (axial capacity). The value of M is sufficient to compare with the component of the resultant external bending moment along this axis for equilibrium or safety in design. To obtain M_R (the resultant moment capacity), M_z (the moment perpendicular to the capacity axis) should be included such that

$$M_R = \left\{M^2 + M_z^2\right\}^{1/2}$$

(2.152)

in which

$$M_z = (V_1 z_1 + V_2 z_2 + V_3 z_3) \tag{2.153}$$

$$V_1 z_1 = 0.50(\cot\theta - \tan\theta)V_1 x_1 + \{z_o - 0.25h(\cot\theta - \tan\theta)\}V_1 \tag{2.154}$$

$$V_2 z_2 = V_2 x_2 \tan\theta \tag{2.155}$$

$$V_3 z_3 = -0.50(\cot\theta - \tan\theta)V_3 x_3 + \{0.25h(\cot\theta - \tan\theta) - z_o\}V_3 \tag{2.156}$$

$$P = (V_1 + V_2 + V_3) \tag{2.157}$$

Note: When $\alpha < \theta < (\pi/2 - \alpha)$, the value of K_2 in Equation 2.143 is changed to

$$K_2 = (h\tan\theta - 2z_o) \tag{2.158}$$

in which

$$\alpha = \arctan(b/d) \tag{2.159}$$

for the outer rectangle, and

$$\alpha = \arctan(b_1/d_1) \tag{2.160}$$

for the inner rectangle. When $\theta > (\pi/2 - \alpha)$, Equation 2.155 is changed to

$$V_2 z_2 = V_2 x_2 / \tan\theta \tag{2.161}$$

and Equation 2.143 is changed to

$$K_2 = [(h/\tan\theta) + 2z_o] \tag{2.162}$$

Equations 2.144, 2.145, 2.150, and 2.151 are changed to the following:

$$x_2 = (1/2)(b\sin\theta - d\cos\theta) \tag{2.163}$$

$$x_1 = (1/2)(b_1\sin\theta - d_1\cos\theta) \tag{2.164}$$

$$z_o = (1/2)(d\sin\theta - b\cos\theta) \tag{2.165}$$

$$z_{o1} = (1/2)(d_1\sin\theta - b_1\cos\theta) \tag{2.166}$$

The pair of values of P and M_R define the yield capacity of the steel section. The plot of these values will trace the yield capacity curve of the steel section as determined from the above equations for rectangular tubing. For a square steel tubular section, set $b = d$ in all the above equations.

2.3.1.5   Capacity curves

Capacity curves are obtained from Microsoft Excel worksheets where the above-listed equations (2.47 to 2.166) are programmed to solve the capacities for any rectangular tubing. Figure 2.14 is the yield capacity curve of a rectangular steel tubing that is 8 × 4 in. (203.2 × 101.6 mm), for which $t = 0.250$ in. (6.35 mm) and $f_y = 36$ ksi (248 MPa). The figure shows the plot of the yield capacity of this rectangular steel tubing at all envelopes of the compressive depth c of the section using the diagonal of the outer rectangular section as the capacity axis.

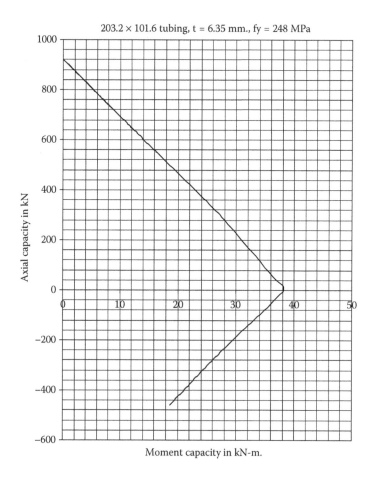

Figure 2.14 Rectangular steel tubing capacity curve.

Note that values below the zero line are indicated for conditions of compressive depth less than the balanced condition. The balanced condition is where the axial capacity is zero and the moment value is a maximum. This is the case for beams subjected to bending loads only.

For sections without tension, the values above the zero line are compared with the external loads the designer is considering. Figure 2.15 shows the column capacity curve for this example of the analytical method in predicting the yield capacity of rectangular tubing for column design.

To use the capacity curve, all that is needed is to determine the external loads to which the pipe will be subjected. Plot the set or sets of external axial and bending moment loads on the capacity curve. Then determine the locations of the plotted values with respect to the capacity curve. When these external loads are within the envelope of the capacity

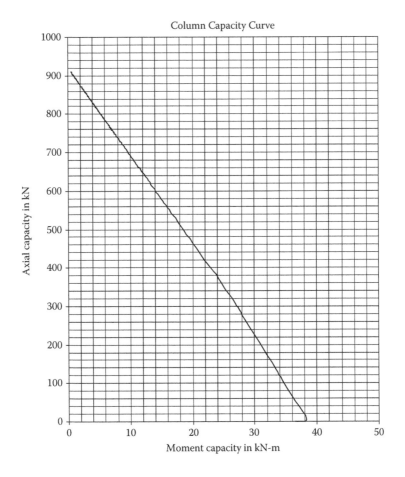

Figure 2.15 Rectangular tubing column capacity curve.

curve, the chosen tubular section is considered sufficient to support the external loads.

The decision as to where inside the curve to limit the external loads is the responsibility of the civil or structural engineer. Every civil/structural engineer can choose all the multipliers to the external loads such as moment magnification factor for slender columns, unanticipated eccentricities of the applied axial load, multipliers to live loads, code requirements, safety factors, and other factors deemed applicable to a particular application.

2.3.1.6   Key points

The schematic locations of the key points on the capacity curve are similar to Figure 2.6. The following descriptions of these key points are for rectangular tubing. Three important key points on the yield capacity curve of rectangular steel tubing are Q, S, and U.

Key point U is the position of the compressive depth c at the center of the rectangular tubing section. It is the so-called balanced condition of loading when the compressive and tensile yield stress of the steel is developed at the same time. The value of the axial capacity is zero.

Key point Q is the condition when the compressive depth c approaches infinity (or a very large number). At that instant, the distribution of compressive stress becomes uniform and the value of the moment capacity approaches zero. The axial capacity is then simply the area of the rectangular tubing times the yield stress of steel.

Key point S is when the compressive depth is equal to h, the start of full compressive stress on the rectangular tubing.

The values of the axial and moment capacities at this point are recommended for minimum eccentricity criteria of the applied external loads.

For columns, key point Q is difficult, if not impossible, to attain because it is inherently out-of-plumb from the vertical axis of any column and because of buckling at mid-height of column length. Because of the unavoidable eccentric loading of external loads on a column section, the minimum axial capacity may be limited at key point S. The value of P at key point S can also be set as the limit of the total compressive force to be applied at the bearing plates for columns.

Key point V is the limit of the capacity curve in the tension zone. This is attained when the compressive depth becomes very small, that is, less than t. When the compressive depth approaches zero, the numerical value of the axial yield capacity approaches that at key point S.

To use the capacity curve, plot sets of external axial and bending moment loads applied to the tubular section. These sets of external loads can represent any combination of loadings required by building codes or any applicable codes in structural design. For safety in design, these sets of points should be inside the capacity curve.

Table 2.8 Values at Key Points

KEY POINT	c inches	MR in-kips	P kips	c cm	MR kN-m	P kN
		theta =	0.4636			
V	0.01	165	-104	0.03	19	-460
U	4.46	340	0	11.33	38	0
S	8.94	179	104	6.60	20	460
Q	c very large	0	207	c very large	0	921

The column capacity curve of the example tubing is shown in Figure 2.15. For columns with predominant tensile capacities, change the label "Compression Zone" to "Tension Zone" and use the same plot of the column capacity curve. The values of the yield capacities at key points are shown in Table 2.8.

2.3.1.7 Capacities of the tubular section at other stresses

Since the steel stress is a multiplier for the integral expressions in Equations 2.47 to 2.151 and can vary up to yield stress in a linear fashion, capacities of the steel tubular section at other stresses can be obtained by proportion as follows:

$$M_1 = (f_s/f_y)M_R \tag{2.167}$$

$$P_1 = (f_s/f_y)P \tag{2.168}$$

From the previous example, when $f_y = 50$ ksi (344.5 MPa), the moment capacity at key point U is from Equation 2.167 and is equal to

$$M_1 = (344.5/248)(38) = 53 \text{ kN-m}$$

Similarly, the capacity at key point Q is from Equation 2.168 and is equal to

$$P_1 = (344.5/248)(921) = 1279 \text{ kN}$$

2.3.1.8 Uniaxial capacities of rectangular tubing

When the position of the capacity axis θ is made equal to zero or $\pi/2$, the uniaxial capacities of the rectangular tubing are obtained. Figure 2.16 is the capacity curve of the example rectangular tubing when $\theta = 0$. In practice, however, the rectangular tubing is always subjected to biaxial bending moments. Hence, the uniaxial yield capacity of a section does not govern the design. It is shown here for information purposes only.

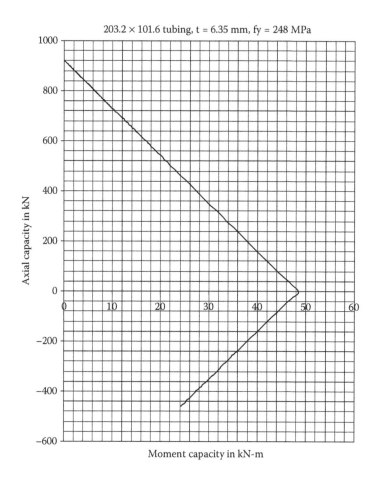

Figure 2.16 Yield capacity of rectangular tubing when $\theta = 0$.

The values of the key points of the uniaxial yield capacity of the rectangular tubing are shown in Table 2.9. Note that the values of M_R in Table 2.9 for uniaxial yield capacity are greater than the values of M_R in Table 2.8 for biaxial yield capacity, that is, 48 kN-m versus 38 kN-m, an increase of 26%. Also, the moment capacities at key point V and S are equal.

Table 2.10 is the uniaxial capacity of the rectangular section when $\theta = \pi/2$. These values are useful for checking the accuracy of the standard interaction formula for biaxial bending. The author recommends using the biaxial yield capacity shown in Table 2.8 for adoption in design criteria for rectangular tubing.

Table 2.9 Values at Key Points

KEY POINT	c inches	MR in-kips	P kips	c cm	MR kN-m	P kN
		theta =	0.0010			
V	0.01	214	-103	0.03	24	-460
U	3.99	427	0	10.14	48	0
S	8.00	214	103	6.60	24	460
Q	c very large	0	207	c very large	0	921

2.3.1.9 Accuracy of the standard interaction formula for biaxial bending

We will use the tabulated values in Tables 2.8, 2.9, and 2.10 to illustrate the serious limitations of the standard interaction formula for biaxial bending. The standard formula was based on the ordinary flexure formula, which is only applicable when the whole section is in compressive stress. When part of the section is in tension, the flexure formula is no longer valid, yet this was the basis for the standard interaction formula for biaxial bending. The standard interaction formula basically states that the sum of the ratios at orthogonal axes should not exceed unity or one.

For our illustration of the limitation of the standard interaction formula for biaxial bending, the capacity of the section at key point S is chosen since, theoretically, the flexure formula is still valid at this point. In this condition of loading, the section is under a compressive triangular loading with the zero stress at key point S. The tabulation is constructed such that $M_1' = M_R' \cos \theta_u$, $M_2' = M_R' \sin \theta_u$, $R_1 = M_1'/M_1$, $R_2 = M_2'/M_2$, $R_3 = P/P_{max}$.

We can see from Table 2.11 that for biaxial bending alone, the standard interaction formula can only capture 80% of the potential capacity of the section. When the axial ratio for column analysis is included in the interaction formula, only 40% can be used by these criteria. From this illustration it is obvious that the standard interaction formula should not be used at all because of its severe limitations in accuracy and correctness.

Table 2.10 Values at Key Points

KEY POINT	c inches	MR in-kips	P kips	c cm	MR kN-m	P kN
		theta =	1.5707			
V	0.01	146	-103	0.03	16	-460
U	1.99	286	0	5.06	32	0
S	4.00	143	103	10.16	16	460
Q	c very large	0	207	c very large	0	921

Table 2.11 Limitations of the Standard Interaction Formula

P =	460	M₁ =	24	Beta =	0.0901				
	MR' =	20		Theta U =	0.3735				
		M2 =	16	PMAX =	921				
M1'	M2'	MR'	R1	R2	R1 + R2	R3	R1 + R2 + R3	R	%
19	7	20	0.776	0.456	1.232	0.499	1.731	1.000	100
18	7	19	0.737	0.433	1.170	0.499	1.670	0.950	95
17	7	18	0.698	0.410	1.109	0.499	1.608	0.900	90
16	6	17	0.659	0.388	1.047	0.499	1.547	0.850	85
15	**6**	**16**	**0.621**	**0.365**	**0.986**	**0.499**	**1.485**	**0.800**	**80**
14	5	15	0.582	0.342	0.924	0.499	1.423	0.750	75
13	5	14	0.543	0.319	0.862	0.499	1.362	0.700	70
12	5	13	0.504	0.296	0.801	0.499	1.300	0.650	65
11	4	12	0.466	0.274	0.739	0.499	1.239	0.600	60
10	4	11	0.427	0.251	0.678	0.499	1.177	0.550	55
9	4	10	0.388	0.228	0.616	0.499	1.115	0.500	50
8	3	9	0.349	0.205	0.554	0.499	1.054	0.450	45
7	**3**	**8**	**0.310**	**0.182**	**0.493**	**0.499**	**0.992**	**0.400**	**40**
7	3	7	0.272	0.160	0.431	0.499	0.931	0.350	35
6	2	6	0.233	0.137	0.370	0.499	0.869	0.300	30
5	2	5	0.194	0.114	0.308	0.499	0.807	0.250	25
4	1	4	0.155	0.091	0.246	0.499	0.746	0.200	20
3	1	3	0.116	0.068	0.185	0.499	0.684	0.150	15

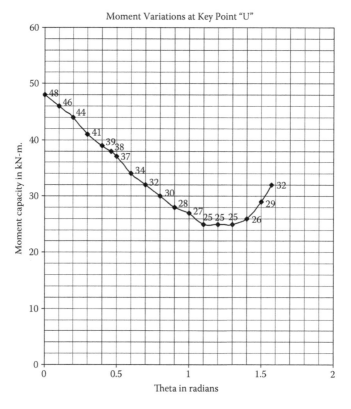

Figure 2.17 Moment capacity versus θ.

2.3.1.10 Variations of moments versus θ

Figure 2.17 shows the variations of bending moment capacity with respect to the inclination of the capacity axis θ. The data used are those obtained in the Excel Worksheets when θ is varied from zero to π/2. The values plotted are those for key point U (balanced condition of loading). From here we see the relative position of the value of the bending moment capacity when θ is along the diagonal of the rectangular tubing. It shows that using the uniaxial capacity at θ = 0 will be unsafe for biaxial bending since this capacity cannot be developed in the section to resist biaxial bending resulting from the application of external loads to the section.

2.3.1.11 Capacity tables for rectangular tubing

Reference capacity Tables 2.12, 2.13, and 2.14 show calculated values of commercially available rectangular steel tubing in English and system

Table 2.12 Capacities of Square Tubing

		SQUARE TUBING CAPACITIES												
		ENGLISH UNITS (M, in IN-KIPS ; P, in KIPS)						SI UNITS (M, in kN-m; P, in kN)						
fy, ksi =	36	KEY POINT Q		KEY POINT S		KEY POINT U		KEY POINT Q		KEY POINT S		KEY POINT U	
size	t	MR	P	MR	P	MR	P	MR	P	MR	P	MR	P
2X2	0.1875	0	49	10	24	19	0	0	218	1	109	2	0
	0.2500	0	63	12	31	23	0	0	280	1	140	3	0
3X3	0.1875	0	76	24	38	47	0	0	338	3	169	5	0
	0.2500	0	99	30	49	59	0	0	440	3	220	7	0
3.5X3.5	0.1875	0	89	33	45	66	0	0	398	4	199	7	0
	0.2500	0	117	42	58	83	0	0	520	5	260	9	0
4X4	0.1875	0	103	44	51	88	0	0	458	5	229	10	0
	0.2500	0	135	56	67	112	0	0	600	6	300	13	0
	0.3125	0	166	67	83	133	0	0	738	8	369	15	0
	0.3750	0	196	77	98	153	0	0	871	9	435	17	0
	0.5000	0	252	93	126	185	0	0	1121	10	560	21	0
5X5	0.1875	0	130	71	65	142	0	0	578	8	289	16	0
	0.2500	0	171	91	85	182	0	0	761	10	380	21	0
	0.3125	0	211	110	105	219	0	0	938	12	469	25	0
	0.3750	0	250	127	125	253	0	0	1111	14	555	29	0
	0.5000	0	324	157	162	312	0	0	1441	18	721	35	0
6X6	0.1875	0	157	104	78	208	0	0	698	12	349	24	0
	0.2500	0	207	135	103	269	0	0	921	15	460	30	0
	0.3125	0	256	163	128	325	0	0	1138	18	569	37	0
	0.3750	0	304	190	152	378	0	0	1351	21	676	43	0
	0.5000	0	396	237	198	473	0	0	1761	27	881	53	0
7X7	0.1875	0	184	144	92	287	0	0	818	16	409	32	0
	0.2500	0	243	187	121	373	0	0	1081	21	540	42	0
	0.3125	0	301	227	150	453	0	0	1339	26	669	51	0
	0.3750	0	358	265	179	529	0	0	1591	30	796	60	0
	0.5000	0	468	335	234	668	0	0	2082	38	1041	76	0
8X8	0.1875	0	211	190	105	379	0	0	938	21	469	43	0
	0.2500	0	279	247	139	493	0	0	1241	28	620	56	0
	0.3125	0	346	302	173	602	0	0	1539	34	769	68	0
	0.3750	0	412	354	206	706	0	0	1831	40	916	80	0
	0.5000	0	540	449	270	897	0	0	2402	51	1201	101	0
	0.6250	0	664	536	332	1069	0	0	2952	61	1476	121	0
10X10	0.1875	0	265	301	132	601	0	0	1178	34	589	68	0
	0.2500	0	351	393	175	786	0	0	1561	44	781	89	0
	0.3125	0	436	483	218	964	0	0	1939	55	970	109	0
	0.3750	0	520	568	260	1135	0	0	2312	64	1156	128	0
	0.5000	0	684	730	342	1457	0	0	3042	82	1521	165	0
	0.6250	0	844	878	422	1753	0	0	3753	99	1876	198	0

Table 2.13 Capacities of Rectangular Tubing

		RECTANGULAR TUBING CAPACITIES										
		ENGLISH UNITS (MR, in IN-KIPS ; P, in KIPS)						SI UNITS (MR, in kN-m; P, in kN)				
fy, ksi =	36	KEY POINT Q		KEY POINT S		KEY POINT U		KEY POINT Q		KEY POINT S		KEY POINT U
size	t	MR	P	MR	P	MR	P	MR	P	MR	P	MR	P
3X2	0.1875	0	62	18	31	36	0	0	278	2	139	4	0
	0.2500	0	81	23	41	44	0	0	360	3	180	5	0
4X2	0.1875	0	76	32	38	60	0	0	338	4	169	7	0
	0.2500	0	99	40	50	75	0	0	440	4	220	8	0
4X3	0.1875	0	89	36	45	72	0	0	398	4	199	8	0
	0.2500	0	117	46	58	90	0	0	520	5	260	10	0
5X2	0.1875	0	89	47	45	89	0	0	398	5	199	10	0
	0.2500	0	117	60	59	113	0	0	520	7	260	13	0
5X3	0.1875	0	103	54	51	105	0	0	458	6	229	12	0
	0.2500	0	135	69	68	133	0	0	600	8	300	15	0
	0.3125	0	166	83	83	159	0	0	738	9	369	18	0
	0.3750	0	196	94	98	181	0	0	871	11	435	21	0
	0.5000	0	252	115	126	220	0	0	1121	13	560	25	0
6X2	0.1875	0	103	65	51	124	0	0	458	7	229	14	0
	0.2500	0	135	83	68	159	0	0	600	9	300	18	0
	0.3125	0	166	99	83	190	0	0	738	11	369	21	0
	0.3750	0	196	114	98	218	0	0	871	13	435	25	0
	0.5000	0	252	139	126	266	0	0	1121	16	560	30	0
6X3	0.1875	0	116	76	58	143	0	0	518	9	259	16	0
	0.2500	0	153	97	77	183	0	0	681	11	340	21	0
	0.3125	0	188	116	94	220	0	0	838	13	419	25	0
	0.3750	0	223	134	111	253	0	0	991	15	495	29	0
	0.5000	0	288	164	144	311	0	0	1281	19	641	35	0
6X4	0.1875	0	130	83	65	161	0	0	578	9	289	18	0
	0.2500	0	171	106	85	207	0	0	761	12	380	23	0
	0.3125	0	211	128	105	249	0	0	938	14	469	28	0
	0.3750	0	250	148	125	288	0	0	1111	17	555	33	0
	0.5000	0	324	183	162	356	0	0	1441	21	721	40	0
7X5	0.1875	0	157	117	78	230	0	0	698	13	349	26	0
	0.2500	0	207	151	104	298	0	0	921	17	460	34	0
	0.3125	0	256	183	128	360	0	0	1138	21	569	41	0
	0.3750	0	304	213	152	419	0	0	1351	24	676	47	0
	0.5000	0	396	267	198	524	0	0	1761	30	881	59	0
8X2	0.1875	0	130	108	65	209	0	0	578	12	289	24	0
	0.2500	0	171	139	85	269	0	0	761	16	380	30	0
	0.3125	0	211	168	105	325	0	0	938	19	469	37	0
	0.3750	0	250	194	125	377	0	0	1111	22	555	43	0
8X3	0.1875	0	143	125	72	237	0	0	638	14	319	27	0
	0.2500	0	189	162	94	307	0	0	841	18	420	35	0
	0.3125	0	233	195	117	371	0	0	1038	22	519	42	0
	0.3750	0	277	227	138	431	0	0	1231	26	615	49	0
	0.5000	0	360	284	180	539	0	0	1601	32	801	61	0
8X4	0.1875	0	157	139	78	263	0	0	698	16	349	30	0
	0.2500	0	207	179	104	340	0	0	921	20	460	38	0
	0.3125	0	256	217	128	412	0	0	1138	25	569	47	0
	0.3750	0	304	253	152	479	0	0	1351	29	676	54	0
	0.5000	0	396	317	198	601	0	0	1761	36	881	68	0

international (SI) units. These values are useful for quick reference in the design of steel tubing. All that is required is to compare the external loads that the civil or structural engineer has determined to be acting on the steel tubing in question. The use of these tables will greatly simplify the task of determining the adequacy and safety of any rectangular steel tubing used in a design.

Table 2.14 Capacities of Rectangular Tubing

		RECTANGULAR TUBING CAPACITIES											
		ENGLISH UNITS (MR, in IN-KIPS ; P, in KIPS)						SI UNITS (MR, in kN-m; P, in kN)					
fy, ksi =	36	KEY POINT Q		KEY POINT S		KEY POINT U		KEY POINT Q		KEY POINT S		KEY POINT U	
size	t	MR	P	MR	P	MR	P	MR	P	MR	P	MR	P
8X6	0.1875	0	184	158	92	312	0	0	818	18	409	35	0
	0.2500	0	243	205	122	404	0	0	1081	23	540	46	0
	0.3125	0	301	249	150	492	0	0	1399	28	669	56	0
	0.3750	0	358	291	179	574	0	0	1591	33	796	65	0
	0.5000	0	468	367	234	725	0	0	2082	41	1041	82	0
10X2	0.1875	0	157	159	78	312	0	0	698	18	349	35	0
	0.2500	0	207	206	104	404	0	0	921	23	460	46	0
	0.3125	0	256	251	128	491	0	0	1138	28	569	55	0
	0.3750	0	304	292	152	572	0	0	1351	33	676	65	0
10X4	0.1875	0	184	204	92	386	0	0	818	23	409	44	0
	0.2500	0	243	265	122	502	0	0	1081	30	540	57	0
	0.3125	0	301	323	150	612	0	0	1338	37	669	69	0
	0.3750	0	358	378	179	716	0	0	1591	43	796	81	0
	0.5000	0	468	479	234	908	0	0	2082	54	1041	103	0
10X6	0.1875	0	211	233	105	449	0	0	938	26	469	51	0
	0.2500	0	279	304	140	585	0	0	1241	34	620	66	0
	0.3125	0	346	371	173	715	0	0	1539	42	769	81	0
	0.3750	0	412	435	206	838	0	0	1831	49	916	95	0
	0.5000	0	540	554	270	1066	0	0	2402	63	1201	120	0
10X8	0.1875	0	238	257	119	511	0	0	1058	29	529	58	0
	0.2500	0	315	335	158	667	0	0	1401	38	701	75	0
	0.3125	0	391	411	195	816	0	0	1739	46	869	92	0
	0.3750	0	466	482	233	959	0	0	2072	55	1036	108	0
	0.5000	0	612	617	306	1225	0	0	2722	70	1361	138	0
12X2	0.1875	0	184	220	92	433	0	0	818	25	409	49	0
	0.2500	0	243	286	122	564	0	0	1081	32	540	64	0
	0.3125	0	301	349	150	687	0	0	1338	39	669	78	0
	0.3750	0	358	408	179	804	0	0	1591	46	796	91	0
12X4	0.1875	0	211	279	105	532	0	0	938	31	469	60	0
	0.2500	0	279	363	140	694	0	0	1241	41	620	78	0
	0.3125	0	346	404	173	848	0	0	1539	50	769	96	0
	0.3750	0	412	522	206	996	0	0	1831	59	916	113	0
	0.5000	0	540	665	270	1271	0	0	2402	75	1201	144	0
12X6	0.1875	0	238	322	119	610	0	0	1058	36	529	69	0
	0.2500	0	315	420	158	796	0	0	1401	47	701	90	0
	0.3125	0	391	515	195	976	0	0	1739	58	869	110	0
	0.3750	0	466	605	233	1147	0	0	2072	68	1036	130	0
	0.5000	0	612	774	306	1468	0	0	2722	87	1361	166	0
12X8	0.1875	0	265	351	132	684	0	0	1178	40	589	77	0
	0.2500	0	351	459	176	895	0	0	1561	52	781	101	0
	0.3125	0	436	563	218	1098	0	0	1939	64	970	124	0
	0.3750	0	520	663	260	1293	0	0	2312	75	1156	146	0
	0.5000	0	684	851	342	1660	0	0	3042	96	1521	188	0

2.4 Steel I-sections

Using the equations in this chapter and the principle of superposition, the yield capacities of steel I-sections can be obtained. To obtain the forces and moments for the flanges of the I-section, use the equations for the outer and inner rectangular sections, that is, Equations 2.47 to 2.151. The same set of equations is then applied to the web of the I-section. The yield capacities of the flanges are the differences between the outer and inner rectangular sections. To this is added the yield capacities of the web section to obtain the yield capacities of the steel I-section.

This section illustrates the analytical method of predicting the yield capacity of steel I-sections subjected to axial and biaxial bending loads. It is based on the linear stress–strain plot of the steel material when subjected to these loads. The resulting expressions for steel forces and moments produce the yield capacity curve of the steel I-section. The analytical method precludes the use of the interaction formula for biaxial bending. Variables considered are the width b and depth d of the I-section, the flange and web thickness, the yield stress f_y of the steel material, and the position of the capacity axis of the I-section.

The capacity axis of the I-section, θ, has to be considered. The optimum position of this axis for biaxial bending is chosen along the diagonal of the outer rectangular section. Most of the external loads applied to the section are located between the horizontal axis and the diagonal. The equations are programmed in a Microsoft Excel worksheet to plot the yield capacity curve of a steel I-section.

The dimensions of the steel I-section are taken from the *AISC Steel Manual*, 7th edition. Yield capacity curves (in English and SI units) for I-sections listed in the AISC manual are provided in tabulated format for easy reference for civil and structural engineers.

2.4.1 Derivation

Figure 2.18 shows the rectangular areas that define the steel I-section. The column capacity axis is the diagonal of the outer rectangular area. It has an

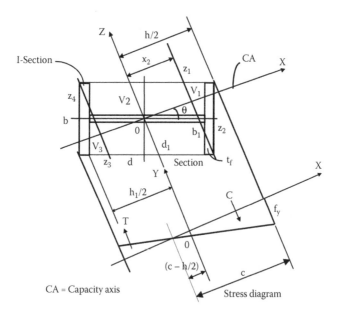

Figure 2.18 Rectangular sections for steel I-section.

inclination of θ from the horizontal axis. The position of θ is shown in its general location.

For every position of this capacity axis, a corresponding set of yield capacity of the steel I-section is obtained using Equations 2.47 to 2.151. We have already shown that the ideal position of this axis should be along the diagonal of the outer rectangle as far as biaxial bending capacity is concerned.

The solution therefore is implemented in the Excel worksheets wherein the equations for a rectangular section are programmed. Three sets of worksheets are required, one each for the outer rectangle, the inner rectangle, and the rectangle defining the web of the I-section.

2.4.1.1 Capacity curves

Capacity curves are plotted from the values predicted by Equation 2.47 to 2.151. Figure 2.19 is the yield capacity curve of the W 16×50 steel I-section, and $f_y = 36$ ksi (248 MPa) of the figure plots the yield capacity of this I-section at all envelopes of the compressive depth c of the section, using the diagonal of the outer rectangular section as the capacity axis.

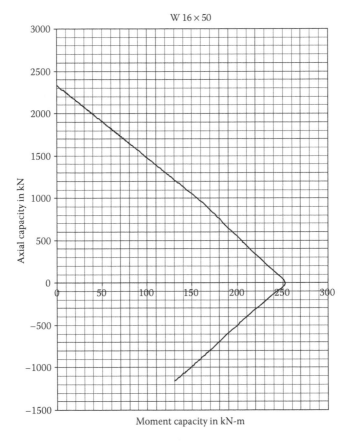

Figure 2.19 Yield capacity curve of I-section.

Note that values below the zero line are indicated for conditions of compressive depth less than the balanced condition. The balanced condition is where the axial capacity is zero and moment value is a maximum. This is the case for beams subjected to bending loads only.

For sections without tension, the values above the zero line are compared with the external loads that a designer considers for his or her design. To use the capacity curve, all that is needed is to determine the external loads to which the I-section will be subjected. Plot the set or sets of external axial and bending moment loads on the capacity curve. Then determine the locations of the plotted values with respect to the capacity curve.

When these external loads are within the envelope of the capacity curve, the chosen I-section is considered sufficient to support the external loads. The decision as to where inside the curve to limit the external loads is the responsibility of the civil or structural engineer. Every civil/structural engineer can choose all the multipliers to the external loads, such as moment magnification factor for slender columns, unanticipated eccentricities of the applied axial load, multipliers to live loads, code requirements, safety factors, and other factors deemed applicable to a particular application.

2.4.1.2   Key points

The schematic locations of the key points on the capacity curve are similar to Figure 2.6. It is sufficient to remember that all the sets of the estimated external axial and bending moments that a given section is subjected to must lie inside the envelope of the capacity curve. Several points could be plotted, such as in bridge design, where several groups of loading are required to be applied in the structural analysis of a steel section for adequacy and safety in design. The values of the yield capacities at key points are shown in Table 2.15 for our example I-section.

2.4.1.3   Capacities of I-sections at other stresses

Since the steel stress is a multiplier for the integral expressions in Equations 2.47 to 2.151 and can vary up to yield stress in a linear fashion, I-section capacities at other stresses can be obtained by using Equations 2.43 and 2.44.

Table 2.15 Capacities at Key Points

KEY POINT	c inches	M_R in-kips	P kips	c cm	M_R kN-m	P kN
		theta =	0.41053			
V	0.01	1147	-262	0.03	130	-1167
U	8.85	2252	-1	22.48	254	0
S	17.72	1206	262	45.02	136	1167
Q	c very large	0	525	c very large	0	2335

From the previous example, when $f_y = 50$ ksi (344.5 MPa), the moment capacity at key point U from Equation 2.43 is equal to

$$M_1 = (344.5/248)(656) = 911 \text{ kN-m}$$

Similarly, the axial capacity at key point Q from Equation 2.44 is equal to

$$P_1 = (344.5/248)(5304) = 7368 \text{ kN}$$

2.4.1.4 Uniaxial capacities of I-sections

When the position of the capacity axis θ is made equal to zero or $\pi/2$, the uniaxial capacities of the I-section are obtained. Figure 2.20 is the capacity curve of the example I-section when $\theta = 0$. In practice, however, the I-section is always subjected to biaxial bending moments. Hence, the uniaxial yield

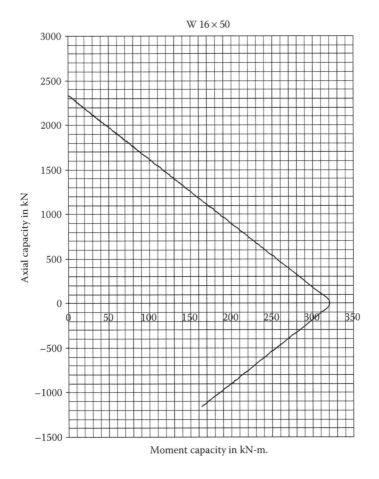

Figure 2.20 Yield capacity curve of I-section when $\theta = 0$.

Table 2.16 Capacities at Key Points

KEY POINT	c inches	MR in-kips	P kips	c cm.	MR kN-m.	P kN
		theta =	0.001			
V	0.01	1437	-262	0.03	162	-1167
U	8.12	2871	0	20.62	325	0
S	16.26	1437	262	41.29	162	1167
Q	c very large	0	525	c very large	0	2335

capacity of a section does not govern the design and is shown here for information purposes only. The values of the key points of the uniaxial yield capacity of the I-section are shown in Table 2.16.

Note that the values of M for uniaxial yield capacity are greater than the values of M in Table 2.15 for biaxial yield capacity. The moment capacity at $\theta = 0$ is 28% more than the moment capacity when θ is chosen at the diagonal of the outer rectangle of the steel I-section in this example.

The author recommends using the biaxial yield capacity shown in Table 2.15 for design criteria for I-sections. This is the maximum moment capacity for the I-section to resist external biaxial bending loads.

2.4.1.5   Variations of moments versus θ

Figure 2.21 plots the variations of bending moment capacity with respect to the inclination of the capacity axis measured by the angle θ.

2.4.1.6   Limitations of the standard interaction formula

Table 2.17 shows the values of the W 16 × 50 steel I-section when $\theta = \pi/2$. Work these values with those in Tables 2.15 and 2.16 to show the limitations

Table 2.17 Values at Key Points

KEY POINT	c inches	MR in-kips	P	c cm	MR kN-m	P kN
		theta =	1.5707			
V	0.01	189	-262	0.03	21	-1167
U	3.53	378	0	8.96	43	0
S	7.07	189	262	17.97	21	1167
Q	c very large	0	525	c very large	0	2335

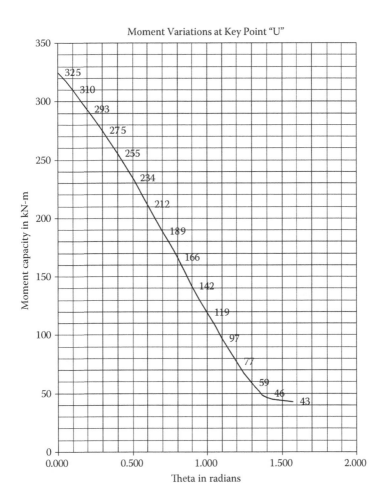

Figure 2.21 Moment capacity versus θ.

and crudeness of the standard interaction formula for biaxial bending. Table 2.18 is constructed such that $M'_1 = M'_R \cos \theta_u$, $M_2 = M'_R \sin \theta_u$, $R_1 = M'_1 / M_1$, $R_2 = M'_2 / M_2$, $R = R_1 + R_2 + R_3$, $\beta = \arctan(M_z/M)$, and $\theta_u = \theta - \beta$. Table 2.18 indicates that the standard interaction formula can only capture 39% of the capacity of the W 16×50 steel section.

2.4.1.7 Capacity tables for I-sections

From the Excel worksheets, we can extract tabulated values of the steel I-sections whose dimensions are taken from the AISC (American Institute

Table 2.18 Limitations of the Standard Interaction Formula for Biaxial Bending

P =	0	MR =	254	Beta =	0.103627	PMAX =	2335		
M =	250	M1 =	325	Theta =	0.4105				
Mz =	26	M2 =	43	Theta U =	0.306873				
M1'	M2'	MR'	R1	R2	R1 + R2	R3	R1 + R2 + R3	R	%
242	77	254	0.745	1.784	2.529	0.000	2.529	1.000	100
238	76	250	0.733	1.756	2.490	0.000	2.490	0.984	98
229	72	240	0.704	1.686	2.390	0.000	2.390	0.945	94
219	69	230	0.675	1.616	2.290	0.000	2.290	0.906	91
210	66	220	0.645	1.546	2.191	0.000	2.191	0.866	87
200	63	210	0.616	1.475	2.091	0.000	2.091	0.827	83
191	60	200	0.587	1.405	1.992	0.000	1.992	0.787	79
181	57	190	0.557	1.335	1.892	0.000	1.892	0.748	75
172	54	180	0.528	1.265	1.792	0.000	1.792	0.709	71
162	51	170	0.499	1.194	1.693	0.000	1.693	0.669	67
153	48	160	0.469	1.124	1.593	0.000	1.593	0.630	63
143	45	150	0.440	1.054	1.494	0.000	1.494	0.591	59
133	42	140	0.411	0.984	1.394	0.000	1.394	0.551	55
124	39	130	0.381	0.913	1.295	0.000	1.295	0.512	51
114	36	120	0.352	0.843	1.195	0.000	1.195	0.472	47
105	33	110	0.323	0.773	1.095	0.000	1.095	0.433	43
95	**30**	**100**	**0.293**	**0.703**	**0.996**	**0.000**	**0.996**	**0.394**	**39**
86	27	90	0.264	0.632	0.896	0.000	0.896	0.354	35
76	24	80	0.235	0.562	0.797	0.000	0.797	0.315	31
67	21	70	0.205	0.492	0.697	0.000	0.697	0.276	28
57	18	60	0.176	0.422	0.597	0.000	0.597	0.236	24
48	15	50	0.147	0.351	0.498	0.000	0.498	0.197	20

of Steel Construction) manual. Tables 2.19 to 2.24 are values of the bending moments and axial capacities of these sections in English units, while Tables 2.25 to 2.30 are values expressed in SI units. These tables are very useful references for any designer using these steel I-sections from the AISC steel manual.

To test the discrepancy of these values as they relate to the sectional areas shown in the AISC steel manual, we can use the proportionality of the change in value as a function of the area of the steel I-section. Using W 16 × 50 in our example, the area by the analytical method is as follows:

Flanges area = 2 × 7.073 × 0.628 = 8.884 in.^{2}
Web area = [16.25 − 2 × 0.628](0.380) = 5.698 in.^{2}
Total area = 14.58 in.^{2} or 14.6 in.^{2}

The area in the AISC steel manual is 14.7 in.^{2} > 14.6 in.^{2}. This indicates that the actual capacities predicted by the analytical method are less than what they should be because the actual area of the section is larger by 0.10 in.^{2}.

To get an idea of the magnitude of this discrepancy, select another I-section next to W 16 × 50. Select W 16 × 45. At key point U the moment capacity of W 16 × 50 is 254 kN-m, while the moment capacity for W 16 × 45 is 228

Table 2.19 Capacity Table for I-Sections

				W-SHAPE CAPACITIES					
				ENGLISH UNITS (M$_R$ in IN-KIPS, P in KIPS)					
f$_y$ = 36 KSI	inches	inches	inches	inches	KEY POINT Q		KEY POINT S		KEY POINT U	
SIZE	d	b	tf	tw	M$_R$	P	M$_R$	P	M$_R$	P
W36X300	36.72	16.655	1.680	0.945	0	3149	16303	1575	30412	0
X280	36.50	16.595	1.570	0.885	0	2939	15190	1469	28326	0
X260	36.24	16.551	1.441	0.841	0	2727	13970	1364	26035	0
X245	36.06	16.512	1.350	0.802	0	2568	13094	1284	24393	0
X230	35.88	16.471	1.260	0.761	0	2408	12224	1204	22764	0
W36X194	36.48	12.117	1.260	0.770	0	2041	10642	1020	20307	0
X182	36.32	12.072	1.180	0.725	0	1912	9955	956	18993	0
X170	36.16	12.027	1.100	0.680	0	1784	9271	892	17687	0
X160	36.00	12.000	1.020	0.653	0	1680	8651	840	16500	0
X150	35.84	11.972	0.940	0.625	0	1574	8030	787	15311	0
X135	35.55	11.945	0.794	0.598	0	1414	6982	707	13306	0
W33X240	33.50	15.865	1.400	0.830	0	2517	11763	1258	21879	0
X220	33.25	15.810	1.275	0.775	0	2308	10714	1154	19916	0
X200	33.00	15.750	1.150	0.715	0	2094	9662	1047	17950	0
W33X152	33.50	11.565	1.055	0.635	0	1596	7732	798	14708	0
X141	33.31	11.535	0.960	0.605	0	1481	7090	740	13484	0
X130	33.10	11.510	0.855	0.580	0	1364	6409	682	12181	0
X118	32.86	11.484	0.738	0.554	0	1236	5656	618	10745	0
W30X210	30.38	15.105	1.315	0.775	0	2204	9245	1102	17159	0
X190	30.12	15.040	1.185	0.710	0	1992	8313	996	15419	0
X172	29.88	14.985	1.065	0.655	0	1803	7470	902	13846	0
W30X132	30.30	10.551	1.000	0.615	0	1386	6025	693	11453	0
X124	30.16	10.521	0.930	0.585	0	1300	5619	650	10679	0
X116	30.00	10.500	0.850	0.564	0	1217	5187	609	9856	0
X108	29.82	10.484	0.760	0.548	0	1132	4722	566	8967	0
X99	29.64	10.458	0.670	0.522	0	1036	4236	518	8041	0
W27X127	27.31	14.090	1.190	0.725	0	1858	6895	929	12781	0
X160	27.08	14.023	1.075	0.658	0	1676	6203	838	11492	0
X145	26.88	13.965	0.975	0.600	0	1519	5608	759	10382	0
W27X114	27.28	10.070	0.932	0.570	0	1197	4684	599	8864	0
X102	27.07	10.018	0.827	0.518	0	1070	4161	535	7872	0
X94	26.91	9.990	0.747	0.490	0	986	3787	493	7160	0
X84	26.69	9.963	0.636	0.463	0	880	3292	440	6220	0
W24X120	24.31	12.088	0.930	0.556	0	1259	4254	629	7888	0
X110	24.16	12.042	0.855	0.510	0	1153	3896	577	7222	0
X100	24.00	12.000	0.775	0.468	0	1048	3526	524	6532	0
W24X94	24.29	9.061	0.872	0.516	0	988	3456	494	6535	0
X84	24.09	9.015	0.772	0.470	0	883	3065	441	5793	0
X76	23.91	8.985	0.682	0.440	0	798	2732	399	5161	0
X68	23.71	8.961	0.582	0.416	0	713	2378	357	4488	0
W24X61	23.72	7.023	0.591	0.419	0	639	2106	319	4048	0
X55	23.55	7.000	0.503	0.396	0	575	1843	287	3543	0
W21X142	21.46	13.132	1.095	0.659	0	1492	4071	746	7586	0
X127	21.24	13.061	0.985	0.588	0	1334	3635	667	6768	0
X112	21.00	13.000	0.865	0.527	0	1175	3177	588	5912	0
W21X96	21.14	9.038	0.935	0.575	0	1007	2971	504	5561	0
X82	20.86	8.962	0.795	0.499	0	859	2517	430	4709	0

Table 2.20 Capacity Table for I-Section

				W-SHAPE CAPACITIES					
				ENGLISH UNITS (MR in IN-KIPS, P in KIPS)					
fy = 36 KSI	inches	inches	inches	inches	KEY POINT Q		KEY POINT S		KEY POINT U	
SIZE	d	b	tf	tw	MR	P	MR	P	MR	P
W21X73	21.24	8.295	0.740	0.455	0	766	2324	383	4377	0
X68	21.13	8.270	0.685	0.430	0	714	2155	357	4057	0
X62	20.99	8.240	0.615	0.400	0	649	1943	325	3656	0
X55	20.80	8.215	0.522	0.375	0	575	1678	288	3156	0
W21X49	20.82	6.520	0.532	0.368	0	511	1495	256	2864	0
X44	20.66	6.500	0.451	0.348	0	459	1303	229	2494	0
W18X114	18.48	11.833	0.991	0.595	0	1198	2745	599	5139	0
X105	18.32	11.792	0.911	0.554	0	1102	2513	551	4701	0
X96	18.16	11.750	0.831	0.512	0	1007	2282	504	4269	0
W18X85	18.32	8.838	0.911	0.526	0	892	2251	446	4184	0
X77	18.16	8.787	0.831	0.475	0	808	2039	404	3789	0
70	18.00	8.750	0.751	0.438	0	733	1840	367	3416	0
64	17.87	8.715	0.686	0.403	0	670	1676	335	3110	0
W18X60	18.25	7.558	0.695	0.416	0	631	1640	315	3075	0
X55	18.12	7.532	0.630	0.390	0	578	1491	289	2794	0
X50	18.00	7.500	0.570	0.358	0	525	1348	263	2525	0
X45	17.86	7.477	0.499	0.335	0	472	1192	236	2231	0
W18X40	17.90	6.018	0.524	0.316	0	419	1087	209	2071	0
X35	17.71	6.000	0.429	0.298	0	366	917	183	1745	0
W16X96	16.32	11.533	0.875	0.535	0	1007	1928	504	3659	0
X88	16.16	11.502	0.795	0.504	0	923	1746	461	3312	0
W16X78	16.32	8.586	0.875	0.529	0	818	1774	409	3290	0
X71	16.16	8.543	0.795	0.486	0	744	1605	372	2974	0
X64	16.00	8.500	0.715	0.443	0	670	1438	335	2663	0
X58	15.86	8.464	0.645	0.407	0	607	1294	303	2395	0
W16X50	16.25	7.073	0.628	0.380	0	525	1206	262	2252	0
X45	16.12	7.039	0.563	0.346	0	472	1079	236	2015	0
X40	16.00	7.000	0.503	0.307	0	419	959	210	1790	0
X36	15.85	6.992	0.428	0.299	0	377	834	188	1555	0
W16X31	15.84	5.525	0.442	0.275	0	324	741	162	1408	0
X26	15.65	5.500	0.345	0.250	0	271	597	136	1133	0
W14X730	22.44	17.889	4.910	3.069	0	7597	13389	3798	26586	0
X665	21.67	17.646	4.522	2.826	0	6895	11820	3448	23509	0
X605	20.94	17.418	4.157	2.598	0	6244	10426	3122	20765	0
X550	20.26	17.206	3.818	2.386	0	5647	9206	2823	18354	0
X500	19.63	17.008	3.501	2.188	0	5097	8138	2549	16233	0
X455	19.05	16.828	3.213	2.008	0	4601	7213	2300	14392	0
W14X426	18.69	16.695	3.033	1.875	0	4287	6659	2143	13286	0
X398	18.31	16.590	2.843	1.770	0	3970	6102	1985	12172	0
X370	17.94	16.475	2.658	1.655	0	3659	5577	1829	11122	0
X342	17.56	16.365	2.468	1.545	0	3343	5061	1672	10087	0
X314	17.19	16.235	2.283	1.415	0	3032	4573	1516	9108	0
X287	16.81	16.130	2.093	1.310	0	2721	4098	1360	8153	0
X264	16.50	16.025	1.938	1.205	0	2463	3721	1232	7395	0
X246	16.25	15.945	1.813	1.125	0	2258	3427	1129	6805	0
W14X237	16.12	15.910	1.748	1.090	0	2152	3278	1076	6506	0
X228	16.00	15.865	1.688	1.045	0	2053	3141	1026	6231	0

Table 2.21 Capacity Table for I-Section

				W-SHAPE CAPACITIES					
				ENGLISH UNITS (M$_R$ in IN-KIPS, P in KIPS)					
f$_y$ = 36 KSI	inches	inches	inches	inches	KEY POINT Q		KEY POINT S		KEY POINT U	
SIZE	d	b	tf	tw	M$_R$	P	M$_R$	P	M$_R$	P
W14X219	15.87	15.825	1.623	1.005	0	1946	2996	973	5939	0
X211	15.75	15.800	1.563	0.980	0	1850	2866	925	5678	0
X202	15.63	15.750	1.503	0.930	0	1752	2734	876	5412	0
X193	15.50	15.710	1.438	0.890	0	1651	2595	825	5133	0
X184	15.38	15.660	1.378	0.840	0	1557	2466	778	4875	0
W14X176	15.25	15.640	1.313	0.820	0	1464	2335	732	4610	0
X167	15.12	15.600	1.248	0.780	0	1369	2202	684	4343	0
X158	15.00	15.550	1.188	0.730	0	1281	2078	640	4096	0
X150	14.88	15.515	1.128	0.695	0	1198	1959	599	3858	0
X142	14.75	15.500	1.063	0.680	0	1116	1836	558	3611	0
W14X320	16.81	16.710	2.093	1.890	0	2801	4196	1401	8366	0
X136	14.75	14.740	1.063	0.660	0	1268	1819	634	3591	0
X127	14.62	14.690	0.998	0.610	0	1170	1692	585	3338	0
X119	14.50	14.650	0.938	0.570	0	1083	1578	541	3110	0
X111	14.37	14.620	0.873	0.540	0	994	1458	497	2871	0
X103	14.25	14.575	0.813	0.495	0	910	1347	455	2649	0
X95	14.12	14.545	0.748	0.465	0	826	1231	413	2418	0
X87	14.00	14.500	0.688	0.420	0	746	1122	373	2203	0
W14X84	14.18	12.023	0.778	0.451	0	878	1290	439	2522	0
X78	14.06	12.000	0.718	0.428	0	815	1184	407	2315	0
X74	14.19	10.072	0.783	0.450	0	772	1292	386	2455	0
X68	14.06	10.040	0.718	0.418	0	709	1178	354	2239	0
X61	13.91	10.000	0.643	0.378	0	635	1048	317	1991	0
X53	13.94	8.062	0.658	0.370	0	550	1017	275	1887	0
X48	13.81	8.031	0.593	0.339	0	497	913	248	1693	0
X43	13.68	8.000	0.528	0.308	0	444	811	222	1502	0
X38	14.12	6.776	0.513	0.313	0	398	786	199	1458	0
X34	14.00	6.750	0.453	0.287	0	355	695	178	1289	0
X30	13.86	6.733	0.383	0.270	0	313	596	156	1104	0
X26	13.89	5.025	0.418	0.255	0	271	545	136	1031	0
X22	13.72	5.000	0.335	0.230	0	229	446	114	843	0
W12X190	14.38	12.670	1.736	1.060	0	1997	2496	999	4957	0
X161	13.88	12.515	1.486	0.905	0	1689	2060	844	4087	0
X133	13.38	12.365	1.236	0.755	0	1386	1652	693	3272	0
X120	13.12	12.320	1.106	0.710	0	1243	1454	622	2877	0
X106	12.88	12.230	0.986	0.620	0	1093	1272	546	2514	0
X99	12.75	12.192	0.921	0.582	0	1016	1177	508	2325	0
X92	12.62	12.155	0.856	0.545	0	939	1084	470	2139	0
W12X85	12.50	12.105	0.796	0.495	0	863	998	432	1967	0
X79	12.38	12.080	0.736	0.470	0	796	916	398	1804	0
X72	12.25	12.040	0.671	0.430	0	719	828	360	1628	0
X65	12.12	12.000	0.606	0.390	0	642	741	321	1455	0
W12X58	12.19	10.014	0.641	0.359	0	603	789	302	1536	0
X53	12.06	10.000	0.576	0.345	0	550	707	275	1375	0
X50	12.19	8.077	0.641	0.371	0	518	778	259	1461	0
X45	12.06	8.042	0.576	0.336	0	465	694	233	1304	0
X40	11.94	8.000	0.516	0.294	0	413	616	206	1157	0

Table 2.22 Capacity Table for I-Section

fy = 36 KSI	inches	inches	inches	inches	KEY POINT Q		KEY POINT S		KEY POINT U	
					\multicolumn{2}{}{W-SHAPE CAPACITIES}					
					\multicolumn{2}{}{ENGLISH UNITS (MR in IN-KIPS, P in KIPS)}					
SIZE	d	b	tf	tw	MR	P	MR	P	MR	P
W12X36	12.24	6.565	0.540	0.305	0	378	631	189	1169	0
X31	12.09	6.525	0.465	0.265	0	325	540	162	1000	0
X27	11.96	6.497	0.400	0.237	0	282	465	141	859	0
X22	12.31	4.030	0.424	0.260	0	230	405	115	772	0
X19	12.16	4.007	0.349	0.237	0	198	340	99	648	0
X16.5	12.00	4.000	0.269	0.230	0	172	278	86	529	0
X14	11.91	3.968	0.224	0.198	0	146	232	73	443	0
W10X112	11.38	10.415	1.248	0.755	0	1161	1155	580	2293	0
X100	11.12	10.345	1.118	0.685	0	1031	1011	515	2006	0
X89	10.88	10.275	0.998	0.615	0	909	884	455	1750	0
X77	10.62	10.195	0.868	0.535	0	778	751	389	1485	0
X72	10.50	10.170	0.808	0.510	0	720	692	360	1368	0
X65	10.38	10.117	0.748	0.457	0	656	633	328	1249	0
X60	10.25	10.075	0.683	0.415	0	590	571	295	1126	0
X54	10.12	10.028	0.618	0.368	0	524	510	262	1004	0
X49	10.00	10.000	0.558	0.340	0	466	456	233	897	0
X45	10.12	8.022	0.618	0.350	0	469	514	234	996	0
X39	9.94	7.990	0.528	0.318	0	405	434	203	842	0
X33	9.75	7.964	0.433	0.292	0	342	354	171	686	0
X29	10.22	5.799	0.500	0.289	0	305	412	152	764	0
X25	10.08	5.762	0.430	0.252	0	262	352	131	652	0
X21	9.90	5.750	0.340	0.240	0	220	283	110	523	0
X19	10.25	4.020	0.394	0.250	0	199	287	100	541	0
X17	10.12	4.010	0.329	0.240	0	177	246	88	462	0
W10X15	10.00	4.000	0.269	0.230	0	156	208	78	391	0
X11.5	9.87	3.950	0.204	0.180	0	119	158	60	296	0
W8X67	9.00	8.287	0.933	0.575	0	695	549	347	1088	0
X58	8.75	8.222	0.808	0.510	0	596	463	298	916	0
X48	8.50	8.117	0.683	0.405	0	489	379	244	749	0
X40	8.25	8.077	0.558	0.365	0	396	303	198	597	0
X35	8.12	8.027	0.493	0.315	0	342	263	171	518	0
X31	8.00	8.000	0.433	0.288	0	296	229	148	449	0
X28	8.06	6.540	0.463	0.285	0	291	248	146	482	0
X24	7.93	6.500	0.398	0.245	0	249	210	125	408	0
X20	8.14	5.268	0.378	0.248	0	209	210	105	392	0
X17	8.00	5.250	0.308	0.230	0	178	172	89	321	0
X15	8.12	4.015	0.314	0.245	0	157	167	78	309	0
X13	8.00	4.000	0.254	0.230	0	135	138	68	255	0
X10	7.90	3.940	0.204	0.170	0	104	108	52	200	0
W6X25	6.37	6.080	0.456	0.320	0	258	147	129	289	0
X20	6.20	6.018	0.367	0.258	0	205	115	102	227	0
X15.5	6.00	5.995	0.269	0.235	0	152	84	76	164	0
X16	6.25	4.030	0.404	0.260	0	168	125	84	235	0
X12	6.00	4.000	0.279	0.230	0	125	87	63	163	0
X8.5	5.83	3.940	0.194	0.170	0	88	60	44	112	0
W5X18.5	5.12	5.025	0.420	0.265	0	177	85	89	167	0
X16	5.00	5.000	0.360	0.240	0	148	71	74	140	0
W4X13	4.16	4.060	0.345	0.280	0	127	46	63	92	0

Table 2.23 Capacity Table for I-Section

fy = 36 KSI	inches	inches	inches	inches	KEY POINT Q		KEY POINT S		KEY POINT U	
					M-SHAPE CAPACITIES					
					ENGLISH UNITS (MR in IN-KIPS, P in KIPS)					
SIZE	d	b	tf	tw	MR	P	MR	P	MR	P
M14X17.2	14.00	4.000	0.272	0.210	0	180	345	90	664	0
M12X11.8	12.00	3.065	0.225	0.177	0	123	199	62	385	0
M10X29.1	9.88	5.937	0.389	0.427	0	306	345	153	639	0
X22.9	9.88	5.752	0.389	0.242	0	240	313	120	579	0
X9	10.00	2.690	0.206	0.157	0	94	128	47	247	0
M8X34.3	8.00	8.003	0.459	0.378	0	330	244	165	480	0
X32.6	8.00	7.940	0.459	0.315	0	320	241	160	474	0
X22.5	8.00	5.395	0.353	0.375	0	236	205	118	384	0
X18.5	8.00	5.250	0.353	0.230	0	194	191	97	357	0
X6.5	8.00	2.281	0.189	0.135	0	68	75	34	145	0
M7X5.5	7.00	2.080	0.180	0.128	0	58	56	29	107	0
M6X22.5	6.00	6.060	0.379	0.372	0	209	113	104	223	0
X20	6.00	5.938	0.379	0.250	0	195	110	97	217	0
X4.4	6.00	1.844	0.171	0.114	0	46	39	23	74	0
M5X18.9	5.00	5.003	0.416	0.316	0	172	81	86	159	0
M4X13.8	4.00	4.000	0.371	0.313	0	124	45	62	90	0
X13	4.00	3.940	0.371	0.254	0	121	45	60	88	0
					HP-SHAPE CAPACITIES					
					ENGLISH UNITS (MR in IN-KIPS, P in KIPS)					
fy = 36 KSI	inches	inches	inches	inches	KEY POINT Q		KEY POINT S		KEY POINT U	
SIZE	d	b	tf	tw	MR	P	MR	P	MR	P
HP14X117	14.23	14.885	0.805	0.805	0	991	1382	495	2714	0
X102	14.03	14.784	0.704	0.704	0	853	1196	426	2340	0
X89	13.86	14.696	0.616	0.616	0	736	1037	368	2024	0
X73	13.64	14.586	0.506	0.506	0	594	842	297	1637	0
HP12X74	12.12	12.217	0.607	0.607	0	703	768	351	1507	0
X53	11.78	12.046	0.436	0.436	0	490	541	245	1053	0
HP10X57	10.01	10.224	0.564	0.564	0	516	479	258	941	0
X42	9.72	10.078	0.418	0.418	0	370	347	185	678	0
HP8X36	8.03	8.158	0.446	0.446	0	332	244	166	480	0

kN-m. The area indicated in the AISC steel manual for W 16×45 is 13.3 in.^{2}. Hence, the change in area is $14.7 - 13.3 = 1.4$ in.^{2}, while the change in the moment capacity is $254 - 228 = 26$ kN-m. The rate of change is $26/1.4 = 18.57$ kN-m/in.^{2}. Multiply this change by 0.10 in.^{2} to obtain 2 kN-m discrepancy. The percentage error = $(2/254) \times 100\% = 0.70\%$, which is negligible.

From this we can conclude that the analytical method is very accurate and is better than the standard interaction formula for biaxial bending.

Table 2.24 Capacity Table for I-Section

fy = 36 KSI	inches	inches	inches	inches	KEY POINT Q		KEY POINT S		KEY POINT U	
					S-SHAPE CAPACITIES					
					ENGLISH UNITS (MR in IN-KIPS, P in KIPS)					
SIZE	d	b	tf	tw	MR	P	MR	P	MR	P
S24X120	24.00	8.048	1.102	0.798	0	1265	4059	632	7742	0
X105.9	24.00	7.875	1.102	0.625	0	1115	3809	558	7276	0
X100	24.00	7.247	0.871	0.747	0	1053	3261	527	6265	0
X90	24.00	7.124	0.871	0.624	0	947	3076	473	5916	0
X79.9	24.00	7.001	0.871	0.501	0	840	2890	420	5563	0
S20X95	20.00	7.200	0.916	0.800	0	998	2546	499	4831	0
X85	20.00	7.053	0.916	0.,653	0	892	2403	446	4565	0
X75	20.00	6.391	0.789	0.641	0	788	2063	394	3947	0
X65.4	20.00	6.250	0.789	0.500	0	687	1918	343	3675	0
S18X70	18.00	6.251	0.691	0.711	0	736	1636	368	3113	0
X54.7	18.00	6.001	0.691	0.461	0	574	1434	287	2735	0
S15X50	15.00	5.640	0.622	0.550	0	525	1011	262	1911	0
X42.9	15.00	5.501	0.622	0.411	0	450	936	225	1771	0
S12X50	12.00	5.477	0.659	0.687	0	524	745	262	1391	0
X40.8	12.00	5.252	0.659	0.462	0	427	677	213	1265	0
X35	12.00	5.078	0.544	0.428	0	367	577	184	1081	0
X31.8	12.00	5.000	0.544	0.350	0	333	552	167	1035	0
S10X35	10.00	4.944	0.491	0.594	0	368	419	184	778	0
X25.4	10.00	4.661	0.491	0.311	0	266	361	133	672	0
S8X23	8.00	4.171	0.425	0.441	0	241	225	121	418	0
X18.4	8.00	4.001	0.425	0.271	0	192	204	96	379	0
S7X20	7.00	3.860	0.392	0.450	0	210	164	105	304	0
X15.3	7.00	3.662	0.392	0.252	0	160	146	80	271	0
S6X17.25	6.00	3.565	0.359	0.465	0	181	113	90	211	0
X12.5	6.00	3.332	0.359	0.232	0	130	100	65	184	0
S5X14.75	5.00	3.284	0.326	0.494	0	154	73	77	138	0
X10	5.00	3.004	0.326	0.214	0	104	64	52	118	0
S4X9.5	4.00	2.796	0.293	0.326	0	99	39	50	75	0
7.7	4.00	2.663	0.293	0.193	0	80	37	40	69	0
S3X7.5	3.00	2.509	0.260	0.349	0	85	18	43	37	0
X5.7	3.00	2.330	0.260	0.170	0	59	18	29	35	0

Table 2.25 Capacity Table for I-Section

				W-SHAPE CAPACITIES					
				SI UNITS (M_R in kN-m, P in kN)					
f_y = 36 KSI	inches	inches	inches	inches	KEY POINT Q		KEY POINT S		KEY POINT U	
SIZE	d	b	tf	tw	M_R	P	M_R	P	M_R	P
W36X300	36.72	16.655	1.680	0.945	0	14009	1842	7004	3436	0
X280	36.50	16.595	1.570	0.885	0	13072	1716	6536	3200	0
X260	36.24	16.551	1.441	0.841	0	12130	1578	6065	2941	0
X245	36.06	16.512	1.350	0.802	0	11423	1479	5712	2756	0
X230	35.88	16.471	1.260	0.761	0	10712	1381	5356	2572	0
W36X194	36.48	12.117	1.260	0.770	0	9077	1202	4538	2294	0
X182	36.32	12.072	1.180	0.725	0	8505	1125	4252	2146	0
X170	36.16	12.027	1.100	0.680	0	7935	1047	3967	1998	0
X160	36.00	12.000	1.020	0.653	0	7471	977	3735	1864	0
X150	35.84	11.972	0.940	0.625	0	7002	907	3501	1730	0
X135	35.55	11.945	0.794	0.598	0	6289	789	3145	1503	0
W33X240	33.50	15.865	1.400	0.830	0	11193	1329	5597	2472	0
X220	33.25	15.810	1.275	0.775	0	10265	1210	5133	2250	0
X200	33.00	15.750	1.150	0.715	0	9316	1092	4658	2028	0
W33X152	33.50	11.565	1.055	0.635	0	7099	874	3550	1662	0
X141	33.31	11.535	0.960	0.605	0	6587	801	3294	1523	0
X130	33.10	11.510	0.855	0.580	0	6067	724	3033	1376	0
X118	32.86	11.484	0.738	0.554	0	5498	639	2749	1214	0
W30X210	30.38	15.105	1.315	0.775	0	9805	1044	4903	1939	0
X190	30.12	15.040	1.185	0.710	0	8863	939	4431	1742	0
X172	29.88	14.985	1.065	0.655	0	8021	844	4011	1564	0
W30X132	30.30	10.551	1.000	0.615	0	6166	681	3083	1294	0
X124	30.16	10.521	0.930	0.585	0	5785	635	2892	1207	0
X116	30.00	10.500	0.850	0.564	0	5414	586	2707	1114	0
X108	29.82	10.484	0.760	0.548	0	5035	533	2518	1013	0
X99	29.64	10.458	0.670	0.522	0	4609	479	2305	908	0
W27X127	27.31	14.090	1.190	0.725	0	8264	779	4132	1444	0
X160	27.08	14.023	1.075	0.658	0	7454	701	3727	1298	0
X145	26.88	13.965	0.975	0.600	0	6756	634	3378	1173	0
W27X114	27.28	10.070	0.932	0.570	0	5325	529	2663	1001	0
X102	27.07	10.018	0.827	0.518	0	4761	470	2381	889	0
X94	26.91	9.990	0.747	0.490	0	4384	428	2192	809	0
X84	26.69	9.963	0.636	0.463	0	3914	372	1957	703	0
W24X120	24.31	12.088	0.930	0.556	0	5599	481	2800	891	0
X110	24.16	12.042	0.855	0.510	0	5131	440	2565	816	0
X100	24.00	12.000	0.775	0.468	0	4661	398	2330	738	0
W24X94	24.29	9.061	0.872	0.516	0	4393	390	2197	738	0
X84	24.09	9.015	0.772	0.470	0	3926	346	1963	654	0
X76	23.91	8.985	0.682	0.440	0	3551	309	1775	583	0
X68	23.71	8.961	0.582	0.416	0	3172	269	1586	507	0
W24X61	23.72	7.023	0.591	0.419	0	2841	238	1421	457	0
X55	23.55	7.000	0.503	0.396	0	2557	208	1279	400	0
W21X142	21.46	13.132	1.095	0.659	0	6639	460	3319	857	0
X127	21.24	13.061	0.985	0.588	0	5934	411	2967	765	0
X112	21.00	13.000	0.865	0.527	0	5227	359	2614	668	0
W21X96	21.14	9.038	0.935	0.575	0	4481	336	2240	628	0
X82	20.86	8.962	0.795	0.499	0	3822	284	1911	532	0

Table 2.26 Capacity Table for I-Section

				W-SHAPE CAPACITIES					
				SI UNITS (MR in kN-m, P in kN)					
fy = 36 KSI	inches	inches	inches	inches	KEY POINT Q		KEY POINT S		KEY POINT U	
SIZE	d	b	tf	tw	MR	P	MR	P	MR	P
W21X73	21.24	8.295	0.740	0.455	0	3406	263	1703	495	0
X68	21.13	8.270	0.685	0.430	0	3175	243	1587	458	0
X62	20.99	8.240	0.615	0.400	0	2889	219	1444	413	0
X55	20.80	8.215	0.522	0.375	0	2560	190	1280	357	0
W21X49	20.82	6.520	0.532	0.368	0	2275	169	1138	324	0
X44	20.66	6.500	0.451	0.348	0	2040	147	1020	282	0
W18X114	18.48	11.833	0.991	0.595	0	5327	310	2664	581	0
X105	18.32	11.792	0.911	0.554	0	4904	284	2452	531	0
X96	18.16	11.750	0.831	0.512	0	4480	258	2240	482	0
W18X85	18.32	8.838	0.911	0.526	0	3968	254	1984	473	0
X77	18.16	8.787	0.831	0.475	0	3593	230	1797	428	0
70	18.00	8.750	0.751	0.438	0	3262	208	1631	386	0
64	17.87	8.715	0.686	0.403	0	2979	189	1490	351	0
W18X60	18.25	7.558	0.695	0.416	0	2805	185	1403	347	0
X55	18.12	7.532	0.630	0.390	0	2573	168	1286	316	0
X50	18.00	7.500	0.570	0.358	0	2336	152	1168	285	0
X45	17.86	7.477	0.499	0.335	0	2099	135	1050	252	0
W18X40	17.90	6.018	0.524	0.316	0	1863	123	931	234	0
X35	17.71	6.000	0.429	0.298	0	1628	104	814	197	0
W16X96	16.32	11.533	0.875	0.535	0	4480	218	2240	413	0
X88	16.16	11.502	0.795	0.504	0	4104	197	2052	374	0
W16X78	16.32	8.586	0.875	0.529	0	3640	200	1820	372	0
X71	16.16	8.543	0.795	0.486	0	3309	181	1654	336	0
X64	16.00	8.500	0.715	0.443	0	2980	162	1490	301	0
X58	15.86	8.464	0.645	0.407	0	2698	146	1349	271	0
W16X50	16.25	7.073	0.628	0.380	0	2335	136	1167	254	0
X45	16.12	7.039	0.563	0.346	0	2100	122	1050	228	0
X40	16.00	7.000	0.503	0.307	0	1865	108	932	202	0
X36	15.85	6.992	0.428	0.299	0	1676	94	838	176	0
W16X31	15.84	5.525	0.442	0.275	0	1441	84	720	159	0
X26	15.65	5.500	0.345	0.250	0	1207	67	603	128	0
W14X730	22.44	17.889	4.910	3.069	0	33791	1513	16895	3004	0
X665	21.67	17.646	4.522	2.826	0	30671	1335	15335	2656	0
X605	20.94	17.418	4.157	2.598	0	27773	1178	13887	2346	0
X550	20.26	17.206	3.818	2.386	0	25116	1040	12558	2074	0
X500	19.63	17.008	3.501	2.188	0	22673	919	11336	1834	0
X455	19.05	16.828	3.213	2.008	0	20464	815	1-232	1626	0
W14X426	18.69	16.695	3.033	1.875	0	19067	752	9534	1501	0
X398	18.31	16.590	2.843	1.770	0	17659	689	8829	1375	0
X370	17.94	16.475	2.658	1.655	0	16275	630	8137	1257	0
X342	17.56	16.365	2.468	1.545	0	14872	572	7436	1140	0
X314	17.19	16.235	2.283	1.415	0	13486	517	6743	1029	0
X287	16.81	16.130	2.093	1.310	0	12101	463	6051	921	0
X264	16.50	16.025	1.938	1.205	0	10957	420	5479	835	0
X246	16.25	15.945	1.813	1.125	0	10042	387	5021	769	0
W14X237	16.12	15.910	1.748	1.090	0	9572	370	4786	735	0
X228	16.00	15.865	1.688	1.045	0	9130	355	4565	704	0

Table 2.27 Capacity Table for I-Section

					W-SHAPES CAPACITIES					
					SI UNITS (MR in kN-m, P in kN)					
f_y = 36 KSI	inches	inches	inches	inches	KEY POINT Q		KEY POINT S		KEY POINT U	
SIZE	d	b	tf	tw	MR	P	MR	P	MR	P
W14X219	15.87	15.825	1.623	1.005	0	8657	339	4329	671	0
X211	15.75	15.800	1.563	0.980	0	8230	324	4115	641	0
X202	15.63	15.750	1.503	0.930	0	7795	309	3898	611	0
X193	15.50	15.710	1.438	0.890	0	7343	293	3671	580	0
X184	15.38	15.660	1.378	0.840	0	6925	279	3462	551	0
W14X176	15.25	15.640	1.313	0.820	0	6510	264	3255	521	0
X167	15.12	15.600	1.248	0.780	0	6089	249	3045	491	0
X158	15.00	15.550	1.188	0.730	0	5698	235	2849	463	0
X150	14.88	15.515	1.128	0.695	0	5329	221	2665	436	0
X142	14.75	15.500	1.063	0.680	0	4966	207	2483	408	0
W14X320	16.81	16.710	2.093	1.890	0	12461	474	6231	945	0
X136	14.75	14.740	1.063	0.660	0	5641	206	2821	406	0
X127	14.62	14.690	0.998	0.610	0	5204	191	2602	377	0
X119	14.50	14.650	0.938	0.570	0	4816	178	2408	351	0
X111	14.37	14.620	0.873	0.540	0	4422	165	2211	324	0
X103	14.25	14.575	0.813	0.495	0	4047	152	2023	299	0
X95	14.12	14.545	0.748	0.465	0	3675	139	1838	273	0
X87	14.00	14.500	0.688	0.420	0	3318	127	1659	249	0
W14X84	14.18	12.023	0.778	0.451	0	3907	146	1954	285	0
X78	14.06	12.000	0.718	0.428	0	3625	134	1812	262	0
X74	14.19	10.072	0.783	0.450	0	3435	146	1718	277	0
X68	14.06	10.040	0.718	0.418	0	3154	133	1577	253	0
X61	13.91	10.000	0.643	0.378	0	2823	118	1412	225	0
X53	13.94	8.062	0.658	0.370	0	2447	115	1223	213	0
X48	13.81	8.031	0.593	0.339	0	2210	103	1105	191	0
X43	13.68	8.000	0.528	0.308	0	1975	92	988	170	0
X38	14.12	6.776	0.513	0.313	0	1770	89	885	165	0
X34	14.00	6.750	0.453	0.287	0	1581	79	791	146	0
X30	13.86	6.733	0.383	0.270	0	1392	67	696	125	0
X26	13.89	5.025	0.418	0.255	0	1206	62	603	117	0
X22	13.72	5.000	0.335	0.230	0	1017	50	509	95	0
W12X190	14.38	12.670	1.736	1.060	0	8884	282	4442	560	0
X161	13.88	12.515	1.486	0.905	0	7511	233	3756	462	0
X133	13.38	12.365	1.236	0.755	0	6167	187	3083	370	0
X120	13.12	12.320	1.106	0.710	0	5529	164	2765	325	0
X106	12.88	12.230	0.986	0.620	0	4861	144	2431	284	0
X99	12.75	12.192	0.921	0.582	0	4518	133	2259	263	0
X92	12.62	12.155	0.856	0.545	0	4178	122	2089	242	0
W12X85	12.50	12.105	0.796	0.495	0	3841	113	1920	222	0
X79	12.38	12.080	0.736	0.470	0	3541	104	1771	204	0
X72	12.25	12.040	0.671	0.430	0	3199	93	1599	184	0
X65	12.12	12.000	0.606	0.390	0	2857	84	1428	164	0
W12X58	12.19	10.014	0.641	0.359	0	2683	89	1341	174	0
X53	12.06	10.000	0.576	0.345	0	2447	80	1224	155	0
X50	12.19	8.077	0.641	0.371	0	2306	88	1153	165	0
X45	12.06	8.042	0.576	0.336	0	2070	78	1035	147	0
X40	11.94	8.000	0.516	0.294	0	1836	70	918	131	0

Table 2.28 Capacity Table for I-Section

					W-SHAPE CAPACITIES					
					SI UNITS (M$_R$ in kN-m, P in kN)					
f$_y$ = 36 KSI	inches	inches	inches	inches	KEY POINT Q		KEY POINT S		KEY POINT U	
SIZE	d	b	tf	tw	M$_R$	P	M$_R$	P	M$_R$	P
W12X36	12.24	6.565	0.540	0.305	0	1680	71	840	132	0
X31	12.09	6.525	0.465	0.265	0	1445	61	723	113	0
X27	11.96	6.497	0.400	0.237	0	1256	53	628	97	0
X22	12.31	4.030	0.424	0.260	0	1024	46	512	87	0
X19	12.16	4.007	0.349	0.237	0	883	38	441	73	0
X16.5	12.00	4.000	0.269	0.230	0	767	31	383	60	0
X14	11.91	3.968	0.224	0.198	0	648	26	324	50	0
W10X112	11.38	10.415	1.248	0.755	0	5163	130	2582	259	0
X100	11.12	10.345	1.118	0.685	0	4584	114	2292	227	0
X89	10.88	10.275	0.998	0.615	0	4045	100	2023	198	0
X77	10.62	10.195	0.868	0.535	0	3459	85	1730	168	0
X72	10.50	10.170	0.808	0.510	0	3203	78	1601	155	0
X65	10.38	10.117	0.748	0.457	0	2917	71	1458	141	0
X60	10.25	10.075	0.683	0.415	0	2625	64	1312	127	0
X54	10.12	10.028	0.618	0.368	0	2329	58	1165	113	0
X49	10.00	10.000	0.558	0.340	0	2071	52	1036	101	0
X45	10.12	8.022	0.618	0.350	0	2086	58	1043	113	0
X39	9.94	7.990	0.528	0.318	0	1803	49	902	95	0
X33	9.75	7.964	0.433	0.292	0	1520	40	760	78	0
X29	10.22	5.799	0.500	0.289	0	1355	47	678	86	0
X25	10.08	5.762	0.430	0.252	0	1166	40	583	74	0
X21	9.90	5.750	0.340	0.240	0	980	32	490	59	0
X19	10.25	4.020	0.394	0.250	0	886	32	443	61	0
X17	10.12	4.010	0.329	0.240	0	786	28	393	52	0
W10X15	10.00	4.000	0.269	0.230	0	693	24	347	44	0
X11.5	9.87	3.950	0.204	0.180	0	531	18	265	33	0
W8X67	9.00	8.287	0.933	0.575	0	3090	62	1545	123	0
X58	8.75	8.222	0.808	0.510	0	2650	52	1325	104	0
X48	8.50	8.117	0.683	0.405	0	2174	43	1087	85	0
X40	8.25	8.077	0.558	0.365	0	1762	34	881	67	0
X35	8.12	8.027	0.493	0.315	0	1523	30	762	59	0
X31	8.00	8.000	0.433	0.288	0	1318	26	659	51	0
X28	8.06	6.540	0.463	0.285	0	1295	28	648	54	0
X24	7.93	6.500	0.398	0.245	0	1108	24	554	46	0
X20	8.14	5.268	0.378	0.248	0	931	24	4654	44	0
X17	8.00	5.250	0.308	0.230	0	790	19	395	36	0
X15	8.12	4.015	0.314	0.245	0	698	19	349	35	0
X13	8.00	4.000	0.254	0.230	0	601	16	301	29	0
X10	7.90	3.940	0.204	0.170	0	461	12	231	23	0
W6X25	6.37	6.080	0.456	0.320	0	1148	17	574	33	0
X20	6.20	6.018	0.367	0.258	0	910	13	455	26	0
X15.5	6.00	5.995	0.269	0.235	0	677	9	339	18	0
X16	6.25	4.030	0.404	0.260	0	748	14	374	27	0
X12	6.00	4.000	0.279	0.230	0	558	10	279	18	0
X8.5	5.83	3.940	0.194	0.170	0	393	7	196	13	0
W5X18.5	5.12	5.025	0.420	0.265	0	788	10	394	19	0
X16	5.00	5.000	0.360	0.240	0	658	8	329	16	0
W4X13	4.16	4.060	0.345	0.280	0	564	5	282	10	0

Table 2.29 Capacity Table for I-Section

				M-SHAPE CAPACITIES						
				SI UNITS (M_R in kN-m, P in kN)						
f_y = 36 KSI	inches	inches	inches	inches	KEY POINT Q		KEY POINT S		KEY POINT U	
SIZE	d	b	tf	tw	M_R	P	M_R	P	M_R	P
M14X17.2	14.00	4.000	0.272	0.210	0	801	39	400	75	0
M12X11.8	12.00	3.065	0.225	0.177	0	548	22	274	43	0
M10X29.1	9.88	5.937	0.389	0.427	0	1362	39	681	72	0
X22.9	9.88	5.752	0.389	0.242	0	1069	35	535	65	0
X9	10.00	2.690	0.206	0.157	0	419	14	209	28	0
M8X34.3	8.00	8.003	0.459	0.378	0	1468	28	734	54	0
X32.6	8.00	7.940	0.459	0.315	0	1422	27	711	54	0
X22.5	8.00	5.395	0.353	0.375	0	1048	23	524	43	0
X18.5	8.00	5.250	0.353	0.230	0	862	22	431	40	0
X6.5	8.00	2.281	0.189	0.135	0	303	9	151	16	0
M7X5.5	7.00	2.080	0.180	0.128	0	256	6	128	12	0
M6X22.5	6.00	6.060	0.379	0.372	0	928	13	464	25	0
X20	6.00	5.938	0.379	0.250	0	865	12	433	24	0
X4.4	6.00	1.844	0.171	0.114	0	204	4	102	8	0
M5X18.9	5.00	5.003	0.416	0.316	0	765	9	383	18	0
M4X13.8	4.00	4.000	0.371	0.313	0	550	5	269	10	0
X13	4.00	3.940	0.371	0.254	0	538	5	269	10	0
				HP-SHAPE CAPACITIES						
				SI UNITS (M_R in kN-m, P in kN)						
f_y = 36 KSI	inches	inches	inches	inches	KEY POINT Q		KEY POINT S		KEY POINT U	
SIZE	d	b	tf	tw	M_R	P	M_R	P	M_R	P
HP14X117	14.23	14.885	0.805	0.805	0	4407	156	2203	307	0
X102	14.03	14.784	0.704	0.704	0	3792	135	1896	264	0
X89	13.86	14.696	0.616	0.616	0	3275	117	1637	229	0
X73	13.64	14.586	0.506	0.506	0	2640	95	1320	185	0
HP12X74	12.12	12.217	0.607	0.607	0	3126	87	1563	170	0
X53	11.78	12.046	0.436	0.436	0	2178	61	1089	119	0
HP10X57	10.01	10.224	0.564	0.564	0	2296	54	1148	106	0
X42	9.72	10.078	0.418	0.418	0	1648	39	824	77	0
HP8X36	8.03	8.158	0.446	0.446	0	1477	28	739	54	0

Table 2.30 Capacity Table for I-Section

fy = 36 KSI	inches	inches	inches	inches	KEY POINT Q		KEY POINT S		KEY POINT U	
					S-SHAPE CAPACITIES					
					SI UNITS (MR in kN-m, P in kN)					
SIZE	d	b	tf	tw	MR	P	MR	P	MR	P
S24X120	24.00	8.048	1.102	0.798	0	5625	459	2813	875	0
X105.9	24.00	7.875	1.102	0.625	0	4961	430	2480	822	0
X100	24.00	7.247	0.871	0.747	0	4684	368	2342	708	0
X90	24.00	7.124	0.871	0.624	0	4211	348	2106	668	0
X79.9	24.00	7.001	0.871	0.501	0	3739	327	1869	629	0
S20X95	20.00	7.200	0.916	0.800	0	4440	288	2220	546	0
X85	20.00	7.053	0.916	0.,653	0	3969	271	1984	516	0
X75	20.00	6.391	0.789	0.641	0	3506	233	1753	446	0
X65.4	20.00	6.250	0.789	0.500	0	3054	217	1527	415	0
S18X70	18.00	6.251	0.691	0.711	0	3275	185	1638	352	0
X54.7	18.00	6.001	0.691	0.461	0	2555	162	1277	309	0
S15X50	15.00	5.640	0.622	0.550	0	2335	114	1167	216	0
X42.9	15.00	5.501	0.622	0.411	0	2001	106	1001	200	0
S12X50	12.00	5.477	0.659	0.687	0	2331	84	1166	157	0
X40.8	12.00	5.252	0.659	0.462	0	1899	76	949	143	0
X35	12.00	5.078	0.544	0.428	0	1633	65	816	122	0
X31.8	12.00	5.000	0.544	0.350	0	1483	62	741	117	0
S10X35	10.00	4.944	0.491	0.594	0	1635	47	818	88	0
X25.4	10.00	4.661	0.491	0.311	0	1182	41	591	76	0
S8X23	8.00	4.171	0.425	0.441	0	1073	25	536	47	0
X18.4	8.00	4.001	0.425	0.271	0	855	23	427	43	0
S7X20	7.00	3.860	0.392	0.450	0	932	18	466	34	0
X15.3	7.00	3.662	0.392	0.252	0	711	17	355	31	0
S6X17.25	6.00	3.565	0.359	0.465	0	803	13	402	24	0
X12.5	6.00	3.332	0.359	0.232	0	579	11	290	21	0
S5X14.75	5.00	3.284	0.326	0.494	0	687	8	343	16	0
X10	5.00	3.004	0.326	0.214	0	463	7	231	13	0
S4X9.5	4.00	2.796	0.293	0.326	0	441	4	220	8	0
7.7	4.00	2.663	0.293	0.193	0	355	4	178	8	0
S3X7.5	3.00	2.509	0.260	0.349	0	379	2	189	4	0
X5.7	3.00	2.330	0.260	0.170	0	262	2	131	4	0

Notations

For pipe section

c: compressive depth of a steel section

t: thickness of a shell

f_y: yield stress of steel

P: axial capacity of the steel pipe section

M: moment capacity of the steel pipe section

$Q, S, U,$ and V: key points on the capacity curve

R: radius of the outer circle

W: plot of the external axial and bending moment loads

M_a: moment of the compressive force around the centerline of circular section

M_b: moment of the tensile force around the centerline of circular section

V_a: compressive force or stress volume on a pipe section

V_b: tensile force or stress volume on a pipe section

M_1: moment capacity of the steel pipe at other specified steel stresses

P_1: axial capacity of the steel pipe at other specified steel stresses

R_1: radius of the inner circle

For rectangular tubing

b: width of the outer rectangle
c: depth of compression
d: length of the outer rectangle
h: overall depth of the outer rectangle
t: thickness of the shell
b_1: width of the inner rectangle
d_1: length of the inner rectangle
h_1: overall depth of the inner rectangle
f_y: yield capacity of steel
f_s: any steel stress less than yield
V_1, V_2, and V_3: forces or stress volumes on tubular sections
V_1x_1, V_2x_2, V_3x_3: moments of forces around centerlines of tubular sections
M: moment capacity of the steel tubular section parallel to the capacity axis
P: axial capacity of the steel tubular section
Q, S, U, and V: key points on the capacity curve
W: plot of the location of external axial and bending moment loads applied on the section
M_z: moment perpendicular to the capacity axis of the tubular section
M_R: resultant bending moment with respect to the capacity axis of the section
M_1: moment capacity of the steel pipe at other specified steel stresses
P_1: axial capacity of the steel pipe at other specified steel stresses
θ: inclination of the capacity axis with the horizontal

Note: All other alphabets and symbols used in the mathematical derivations are defined in the context of their use.

For I-sections

b: flange width or width of the outer rectangle
c: depth of compression
d: depth of the I-section or length of the outer rectangle
h: overall depth of the outer rectangle
b_1: web thickness
d_1: depth of the web or length of the inner rectangle
h_1: overall depth of the inner rectangle
f_y: yield capacity of steel
f_s: any steel stress less than yield
t_f: flange thickness
M: component of the resultant bending moment capacity of the I-section along the capacity axis
P: axial capacity of the I-section
W: location of the external axial and bending moment loads inside the capacity curve

M_R: resultant bending moment capacity of the I-section

α: arctan(b/d) position of the capacity axis for the plotted capacity charts

θ: inclination of the capacity axis with the horizontal

Note: All other alphabets and symbols used in the mathematical derivations are defined in the context of their use.

chapter three

Reinforced concrete sections

3.1 Introduction

For more than five decades, engineers used the equivalent rectangular stress block, the finite-element methods, and the interaction formula to predict the ultimate strength of reinforced concrete sections subjected to axial and biaxial bending loads. The analytical method using basic mathematics and the known stress–strain properties of concrete and steel materials is illustrated in this chapter and compared to the available literature that employs graphical methods of solution such as the standard interaction formula for biaxial bending.

The analytical method precludes the use of graphical methods and is the only work done on this subject utilizing the true parabolic stress method of analysis in reinforced concrete. The true parabolic stress method utilizes the basic parabola, which closely fits the stress–strain curves of concrete cylinders. This parabola is defined by the value of ultimate concrete compressive stress, f_c', the depth of compression, c, and the useable value of ultimate strain in concrete, e_c. The maximum area derived from these limiting parameters is the measure of the ultimate strength of the concrete section.

The use of the parabola is defined under ACI318.10.2.6 regulations. The author presented the equation of this parabola at the First International Structural Engineering and Construction Conference (ISEC-01) held in Honolulu, Hawaii. The derivation of this parabola is shown in the author's first book, *Analytical Method in Reinforced Concrete* (2004). This basic parabola is part of the stress-strain diagram to solve for the concrete and bar forces in a given reinforced concrete column section in the development of its ultimate strength capacities.

The variables considered are concrete compressive depth, c; concrete useable strain, e_c; ultimate concrete compressive stress, f_c'; steel strain, e_s, and the corresponding steel yield stress, f_y; the inclination of the column capacity axis, θ; the radius, R, of the circular column; the width, b and depth, d, of the rectangular column; and the size, number, spacing, and concrete cover of reinforcing steel bars.

The analytical method uses the classical strength of materials approach, basic calculus, and the fundamental requirement of equilibrium conditions defined by $\Sigma F = 0$ and $\Sigma M = 0$ in any structural analysis. For concrete, it uses the parabolic nature of the concrete stress–strain curve and the useable concrete strain set by codes of practice. For steel, it utilizes the linear property of the steel stress–strain curve. These two material properties are linked to a common deformation as they resist external loads. This deformation is assumed to be linear with respect to the neutral axis defined by the compressive depth, c, of the concrete section. One major difference between the graphical method and the analytical method is the introduction of the column capacity axis, which can vary from 0 to $\pi/2$ from the horizontal axis and defines the ultimate strength of the column section from uniaxial to any position of biaxial bending conditions. This axis effectively does away with the concept of the interaction formula for biaxial bending in the graphical method of analysis. It also allows the determination of the centroid of internal forces, which can correspond to a specific external load, which is lacking in the current graphical method.

The use of the analytical method has also demonstrated the diagonal of the rectangular section as the axis for minimum capacity to biaxial bending, and for the circular section the column capacity axis can lie between any two steel bars for the equilibrium of internal and external forces.

The concrete forces are determined by integration of the stress volumes at every position of the concrete compressive depth, using the true parabola instead of the equivalent rectangular stress block. The value of the concrete forces is not affected by the useable concrete strain assigned by codes. The steel forces, however, are governed by the useable concrete strain, chosen in the analysis, and the limiting condition for the maximum yield stress of steel. The yield stress and strain of a steel bar is the limiting point from the neutral axis to the tensile stress of the farthest bar from this axis. This standard stress–strain distribution is known as the CRSI (Concrete Reinforcing Steel Institute) method, in which the straight-line shape of the strain diagram is pivoted at the position of useable concrete strain equal to 0.003. Other values of the useable concrete strain can be used, such as the Canadian method wherein the concrete strain is assumed to be 0.0035.

For the illustrated examples, the columns' capacity curves and tables generated from Excel spreadsheets are based on the ACI (American Concrete Institute) value of concrete strain equal to 0.003. For a rational comparison of results between the graphical and analytical methods, values from the analytical method are used in the standard interaction formula for biaxial bending. The comparison proved that the analytical method should be adopted and the standard interaction formula for biaxial bending discarded since it is an inaccurate procedure.

The analysis is limited to standard shapes of reinforced concrete such as circular and rectangular sections. The methodology is applicable not only to concrete material, but also to timber and steel, whose stress–strain properties can be expressed in equation form for analysis. Chapter 2 discusses the applicability of the analytical approach for steel sections.

3.2 Stress diagram

Figure 3.1 shows the basic parabola and the stress–strain diagrams for the development of the internal forces in a column section. It shows the CRSI method, which uses a straight line pivoted at the concrete strain value of 0.003 or 0.0035, as desired. As this straight line is pivoted around the assumed limit of concrete strain; it intersects the horizontal line at a point designated as the neutral axis or zero value of stress and strain. The distance of this point from the compressive edge is designated as c. The balanced condition of loading is attained when the steel tensile stress and strain is developed simultaneously with the concrete stress and strain. This condition of loading in the section will produce the section's maximum bending moment capacity.

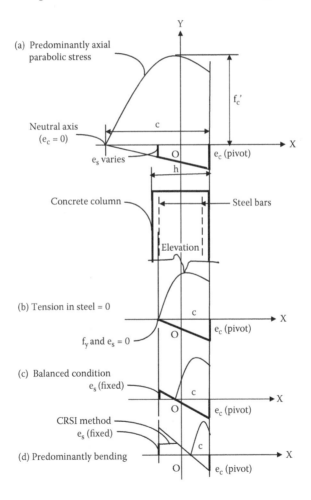

Figure 3.1 Stress–strain diagrams for the analytical method.
Source: Jarquio, Ramon V. *Analytical Method in Reinforced Concrete.* Universal Publishers, Boca Raton, FL. p. 30. Reprinted with Permission.

When c is less than its value at a balanced condition, the tension portion of the section will exceed the yield stress of the steel since a concrete strain of 0.003 is greater than that of the steel strain. Hence, the value of the steel strain is limited to $f_y/29,000$ or $f_y/200,000$ in standard international (SI) units using Hooke's law.

For biaxial analysis, the column capacity axis is the reference axis for determining the ultimate strength of circular and rectangular column sections. Calculations referred to this axis will generate ultimate strength values for every position of c from the concrete compressive edge.

3.2.1 Circular sections

In Figure 3.2 the equation of the basic parabola is written as

$$y = (0.75\, f_c'/c^2)[(c^2 + 2Rc - 3R^2) + 2(3R - c)x - 3x^2] \tag{3.1}$$

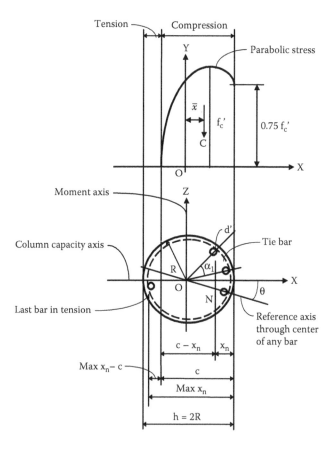

Figure 3.2 Circular column section.
Source: Jarquio, Ramon V. *Analytical Method in Reinforced Concrete.* Universal Publishers, Boca Raton, FL. p. 32. Reprinted with Permission.

The equation of the circle is

$$x^2 + z^2 = R^2 \tag{3.2}$$

From Equations 3.1 and 3.2, the concrete compressive force, C, is given by the expression

$$C = [1.50\,f_c'/c^2] \int_{-(c-R)}^{R} [(c^2 + 2Rc - 3R^2) + 2(3R - c)\,x - 3x^2]\{R^2 - x^2\}^{1/2}\,dx \tag{3.3}$$

Integrate Equation 3.3 using standard integration formulas and evaluate the limits to obtain

$$C = [1.50\,f_c'/c^2][A + B] \tag{3.4}$$

in which

$$A = \{2Rc - c^2\}^{1/2}\,\{0.417\,c^3 + 1.875\,R^3 - [0.375R^2c + 0.583\,Rc^2]\} \tag{3.5}$$

$$B = [\pi R^2/4]\{1 + [\arcsin(c - R)/R][1/90]\}[c^2 + 2Rc - 3.75R^2] \tag{3.6}$$

The above formulas are to be used when $R < c < 2R$.
   When $0 < c \leq R$,

$$A = \{2Rc - c^2\}^{1/2}\{1.917\,c^3 + 1.875\,R^3 + 2.625\,R^2c - 5.083Rc^2\} \tag{3.7}$$

$$B = [\pi R^2/4]\{(1 + [\arcsin(c - R)/R][1/90])[c^2 + 2Rc - 3.75R^2]$$
$$- (3R^2/4)(1 + [\arcsin(c - R)/R][1/90])\} \tag{3.8}$$

The centroid of the compressive force, C, is determined by evaluating the expression

$$C\,\bar{x} = [1.50\,f_c'/c^2] \int_{-(c-R)}^{R} [(c^2 + 2Rc - 3R^2)\,x + 2(3R - c)\,x2 - 3x^3]\{R^2 - x^2\}^{1/2}\,dx$$

$$\tag{3.9}$$

Using standard integration formulas,

$$C\,\bar{x} = [1.50\,f_c'/c^2][D + E] \tag{3.10}$$

$$D = \{2Rc - c^2\}^{1/2}\,\{0.60Rc^3 - [0.75R^4 + 0.233c^4 + 0.017R^2c^2]\} \tag{3.11}$$

$$E = [\pi R^4/8][3R - c]\{1 + [\arcsin(c - R)/R][1/90]\} \tag{3.12}$$

The above formulas are to be used when $R < c < 2R$.
    When $0 < c \leq R$,

$$D = \{2Rc - c^2\}^{1/2}\{0.60Rc^3 - [0.75R^4 + 0.233c^4 + 0.017R^2c^2]\} \qquad (3.13)$$

From the above expressions,

$$\bar{x} = [D + E]/[A + B] \qquad (3.14)$$

Similarly, when $c \geq 2R$,

$$C = [1.50\,f_c'/c^2]\{[\pi R^2/2][c^2 + 2Rc - 3R^2] - [3\pi R^4/8]\} \qquad (3.15)$$

and

$$C\,\bar{x} = (1.50\,f_c'/c^2)[(\pi R^4/4)(3R - c)] \qquad (3.16)$$

Hence,

$$\bar{x} = [2R^2\,(3R - c)]/[4c^2 + 8Rc - 15R^2] \qquad (3.17)$$

3.2.2 Rectangular sections

In Figure 3.3 let b and d denote, respectively, the width and depth of the rectangular column section. From the rotation of axes, or using trigonometry,

$$h/2 = (1/2)[d \cos \theta + b \sin \theta] \qquad (3.18)$$

The equations of the lines representing the four sides of the rectangular section are listed in Chapter 1 as Equations 1.6 to 1.9 and will be utilized with the following expressions such as

$$z_0 = (1/2)[b \cos \theta - d \sin \theta] \quad \text{when } \theta < [(\pi/2) - \alpha];$$
$$z_0 = (1/2)[d \sin \theta - b \cos \theta] \quad \text{when } \theta > [(\pi/2) - \alpha]$$

$$x_2 = (1/2)[d \cos \theta - b \sin \theta] \quad \text{when } \theta \text{ is between zero and } [(\pi/2) - \alpha]$$

and

$$x_2 = (1/2)[b \sin \theta - d \cos \theta] \quad \text{when } \theta \text{ is greater than } [(\pi/2) - \alpha]$$

where $\alpha = \arctan(b/d)$ and $\theta = $ the inclination of the column capacity axis.
    Divide the area of the rectangle into three sections such as V_1, V_2, and V_3. V_1 is the stress volume whose maximum limits are from $-x_2$ to x_2. V_2 is the stress volume whose maximum limits are from x_2 to $h/2$. V_3 is the stress volume whose maximum limits are from $-h/2$ to $-x_2$. Depending upon the position of the compressive concrete depth, c, V_1, V_2, and V_3 will vary and will reach their maximum values indicated in the above limits.

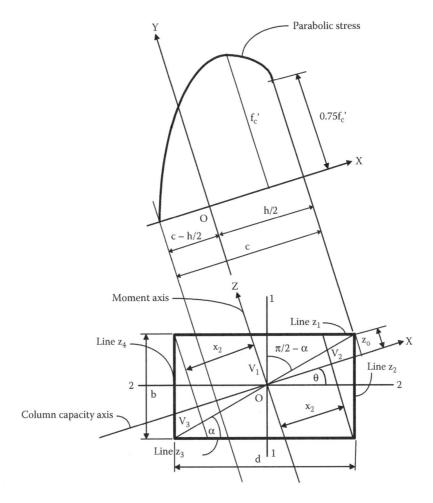

Figure 3.3 Rectangular column with biaxial bending (showing concrete dimensions). *Source:* Jarquio, Ramon V. *Analytical Method in Reinforced Concrete.* Universal Publishers, Boca Raton, FL. p. 52. Reprinted with Permission.

The equation of the basic parabola referred to as the rotated position of X and Y is written as

$$y = [0.75f_c'/c^2][(c^2 + ch - 0.75h^2) + (3h - 2c)\,x - 3x^2] \qquad (3.19)$$

The compressive force on the concrete section is the volume enclosed by the stress diagram and the area of the concrete section defined by c. Hence,

$$V_1 = [0.75\,f_c'/c^2][h\,\tan\,\theta + 2z_0]\int_{-(c-h/2)}^{x2}[(c^2 + ch - 0.75h^2) + (3h - 2c)x - 3x^2]dx$$

$$(3.20)$$

The moment of this force around the centerline (Z-axis) of the column section is given by

$$V_1 \bar{x}_1 = [0.75 f_c'/c^2][h \tan \theta + 2z_0] \int_{-(c-h/2)}^{x2} [(c^2 + ch - 0.75h^2)x$$

$$+ (3h - 2c)x^2 - 3x^3]dx \tag{3.21}$$

$$V_1(\text{max}) = [0.75 f_c'/c^2][h \tan \theta + 2z_0] \int_{-x2}^{x2} [(c^2 + ch - 0.75h^2)$$

$$+ (3h - 2c)x - 3x^2] \, dx \tag{3.22}$$

$$V_1 \bar{x}_1(\text{min}) = [0.75 f_c'/c^2][h \tan \theta + 2z_0] \int_{-x2}^{x2} [(c^2 + ch - 0.75h^2)x$$

$$+ (3h - 2c)x^2 - 3x^3]dx \tag{3.23}$$

$$V_2 = -[0.75 f_c'/c^2][\cot \theta + \tan \theta] \int_{(h/2-c)}^{h/2} [-(0.50c^2h + 0.50ch^2 - 0.375h^3)$$

$$+ (c^2 + 2ch - 2.25h^2)x + (4.5h - 2c)x^2 - 3x^3] \, dx \tag{3.24}$$

$$V_2 \bar{x}_2 = -[0.75 f_c'/c^2][\cot \theta + \tan \theta] \int_{(h/2-c)}^{h/2} [-(0.50c^2h + 0.50ch^2 - 0.375h^3)]x$$

$$+ (c^2 + 2ch - 2.25h^2)x^2 + (4.5h - 2c)x^3 - 3x^4] \, dx \tag{3.25}$$

$$V_2 (\text{max}) = -[0.75 f_c'/c^2][\cot \theta + \tan \theta] \int_{x2}^{h/2} [-(0.50c^2h + 0.50ch^2 - 0.375h^3)]$$

$$+ (c^2 + 2ch - 2.25h^2)x + (4.5h - 2c)x^2 - 3x^3] \, dx \tag{3.26}$$

$$V_2 \bar{x}_2 (\text{max}) = -[0.75 f_c'/c^2][\cot \theta + \tan \theta] \int_{x2}^{h/2} [-(0.50c^2h + 0.50ch^2 - 0.375h^3)]x$$

$$+ (c^2 + 2ch - 2.25h^2)x^2 + (4.5h - 2c)x^3 - 3x^4] \, dx \tag{3.27}$$

$$V_3 = [0.75 f_c'/c^2][\cot \theta + \tan \theta] \int_{-(c-h/2)}^{-x2} [(0.50c^2h + 0.50ch^2 - 0.375h^3)]$$

$$+ (c^2 + 0.75h^2)x + (1.50h - 2c)x^2 - 3x^3] \, dx \tag{3.28}$$

$$V_3 \bar{x}_3 = [0.75 f_c'/c^2][\cot \theta + \tan \theta] \int_{-(c-h/2)}^{-x2} [(0.50c^2h + 0.50ch^2 - 0.375h^3)]x$$

$$+ (c^2 + 0.75h^2)x^2 + (1.50h - 2c)x^3 - 3x^4] \, dx \tag{3.29}$$

$$V_3 \text{ (max)} = [0.75 f_c'/c^2][\cot \theta + \tan \theta] \int_{-h/2}^{-x2} [0.50c^2h + 0.50ch^2 - 0.375h^3]$$

$$+ (c^2 + 0.75h^2)x + (1.50h - 2c)x^2 - 3x^3] \, dx \tag{3.30}$$

$$V_3 \bar{x}_3 \text{ (max)} = [0.75 f_c'/c^2][\cot \theta + \tan \theta] \int_{-h/2}^{-x2} [(0.50c^2h + 0.50ch^2 - 0.375h^3]x$$

$$+ (c^2 + h^2)x^2 + (1.50h - 2c)x^3 - 3x^4] \, dx \tag{3.31}$$

Using standard integration formulas, the following are obtained:

$$V_1 = [0.75 f_c'/c^2][h \tan \theta + 2z_o][i + j + k] \tag{3.32}$$

in which $i = [c^2 + ch - 0.75h^2][x_2 + (c - h/2)$;
$\quad j = [3h - 2c][1/2][x_2^2 - (c - h/2)^2]$;
and $k = -[x_2^3 + (c - h/2)^3]$.

$$V_1 \bar{x}_1 = [0.75 f_c'/c^2][h \tan \theta + 2z_o][i + j + k] \tag{3.33}$$

in which $i = [c^2 + ch - 0.75h^2][1/2][x_2^2 - (c - h/2)^2]$;
$\quad j = [3h - 2c][1/3][x_2^3 + (c - h/2)^3]$;
and $k = -0.75[x_2^4 - (c - h/2)^4]$.

$$V_1 \text{ (max)} = [0.75 f_c'/c^2][h \tan \theta + 2z_o][i + j + k] \tag{3.34}$$

in which $i = [c^2 + ch - 0.75h^2][2x_2]$;
$\quad j = 3h - 2c][1/2][x_2^2 - x_2^2] = 0$;
and $k = [x_2^3 + x_2^3] = -2x_2^3]$.

$$V_1 \bar{x}_1 \text{ (min)} = [0.75 f_c'/c^2][h \tan \theta + 2z_o][i + j + k] \tag{3.35}$$

in which $i = [c^2 + ch - 0.75h^2][1/2][x_2^2 - x_2^2] = 0$;
$\quad j = [3h - 2c][1/3][x_2^3 + x_2^3] = [3h - 2c][2/3]x_2^3$;
and $k = -[x_2^4 - x_2^4] = 0$.

$$V_2 = -[0.75 f_c'/c^2][\cot \theta + \tan \theta][i + j + k + l] \tag{3.36}$$

in which $i = -[0.50c^2h + 0.50ch^2 - 0.375h^3][c]$;
$\quad j = [c^2 + 2ch - 2.25h^2][1/2][(h/2)^2 - (h/2 - c)^2]$;
$\quad k = [4.5h - 2c][1/3][(h/2)^3 - (h/2 - c)^3]$;
and $l = -0.75[(h/2)^4 - (h/2 - c)^4]$.

$$V_2 \bar{x}_2 = -[0.75 f_c'/c^2][\cot \theta + \tan \theta][i + j + k + l] \tag{3.37}$$

in which $i = -[0.50c^2h + 0.50ch^2 - 0.375h^3][1/2][(h/2)^2 - (h/2 - c)^2]$;

$\quad j = [c^2 + 2ch - 2.25h^2][1/3][(h/2)^3 - (h/2 - c)^3]$;

$\quad k = [4.5h - 2c][1/4][(h/2)^4 - (h/2 - c)^4]$;

and $l = -0.60[(h/2)^5 - (h/2 - c)^5]$.

$$V_2 \text{ (max)} = -[0.75f_c' \, f_c'/c^2][\cot \theta + \tan \theta][i + j + k + l] \tag{3.38}$$

in which $i = -[0.50c^2h + 0.50ch^2 - 0.375h^3][(h/2) - x_2]$;

$\quad j = [c^2 + 2\,ch - 2.25h^2] \, [1/2][(h/2)^2 - x_2^2]$;

$\quad k = [4.5h - 2c][1/3][(h/2)^3 - x_2^3]$;

and $l = -0.75[(h/2)^4 - x_2^4]$.

$$V_2 \, \bar{x}_2(\text{max}) = -[0.75 \, f_c'/c^2][\cot \theta + \tan \theta][i + j + k + l] \tag{3.39}$$

in which $i = -[0.50c^2h + 0.50ch^2 - 0.375h^3][1/2][(h/2)^2 - x_2^2]$;

$\quad j = [c^2 + 2ch - 2.25 \, h^2] \, [1/3][(h/2)^3 - x_2^3]$;

$\quad k = [4.5h - 2c][1/4][(h/2)^4 - x_2^4]$;

and $l = -0.60[(h/2)^5 - x_2^5]$.

$$V_3 = [0.75 \, f_c'/c^2][\cot \theta + \tan \theta][i + j + k + l] \tag{3.40}$$

in which $i = [0.50c^2h + 0.50ch^2 - 0.375h^3][(c - h/2) - x_2]$;

$\quad j = -[c^2 + 0.75h^2][1/2][(c - h/2)^2 - x_2^2]$;

$\quad k = [1.50h - 2c][1/3][(c - h/2)^3 - x_2^3]$;

and $l = 0.75[(c - h/2)^4 - x_2^4]$.

$$V_3 \, \bar{x}_3 = [0.75f_c' \, f_c'/c^2][\cot \theta + \tan \theta][i + j + k + l] \tag{3.41}$$

in which $i = -[0.50c^2h + 0.50ch^2 - 0.375h^3][1/2][(c - h/2)^2 - x_2^2]$;

$\quad j = [c^2 + 0.75h^2] \, [1/3][(c - h/2)^3 - x_2^3]$;

$\quad k = -[1.50h - 2c][1/4][(c - h/2)^4 - x_2^4]$;

and $l = -0.60[(c - h/2)^5 - x_2^5]$.

$$V_3 \text{ (max)} = [0.75 \, f_c'/c^2][\cot \theta + \tan \theta][i + j + k + l] \tag{3.42}$$

in which $i = [0.50c^2h + 0.50ch^2 - 0.375h^3][(h/2) - x_2]$;

$\quad j = -[c^2 + 0.75h^2][1/2][(h/2)^2 - x_2^2]$;

$\quad k = [1.50h - 2c][1/3][(h/2)^3 - x_2^3]$;

and $l = 0.75[(h/2)^4 - x_2^4]$.

$$V_3 \, \bar{x}_3 \text{ (max)} = [0.75 \, f_c'/c^2][\cot \theta + \tan \theta][i + j + k + l] \tag{3.43}$$

in which $i = -[0.50c^2h + 0.50ch^2 - 0.375h^3][1/2][(h/2)^2 - x_2^2]$;

$\quad j = [c^2 + 0.75h^2] \, [1/3][(h/2)^3 - x_2^3]$;

$\quad k = -[1.50h - 2c][1/4][(h/2)^4 - x_2^4]$;

and $l = -0.60[(h/2)^5 - x_2^5]$.

Concrete forces are thus obtained using the preceding equations, that is,

$$\text{Concrete force} = \text{Sum}(V_1 + V_2 + V_3) \tag{3.44}$$

when the whole section is in compression. Otherwise, it is the sum of sections in compression as defined by the compressive depth c since concrete is considered to have no resistance to tension. Similarly,

$$\text{Concrete moment} = \text{Sum } (V_1 \bar{x}_1 + V_2 \bar{x}_2 + V_3 \bar{x}_3) \qquad (3.45)$$

There are four envelopes of values for c and the corresponding equations to use for concrete forces:

1. When $0 < c < \{(h/2) - x_2\}$. $V_1 = 0$, $V_3 = 0$, $V_2 =$ Equation 3.36, and $V_2\bar{x}_2 =$ Equation 3.37.
2. When $\{(h/2) - x_2\} < c < \{(h/2) + x_2\}$. $V_3 = 0$, $V_1 =$ Equation 3.32, $V_1\bar{x}_1 =$ Equation 3.33, $V_2 =$ Equation 3.38, and $V_2\bar{x}_2 =$ Equation 3.39.
3. When $\{(h/2) + x_2\} < c < h$. $V_1 =$ Equation 3.34, $V_1\bar{x}_1 =$ Equation 3.35, $V_2 =$ Equation 3.38, $V_2\bar{x}_2 =$ Equation 3.39, $V_3 =$ Equation 3.40, and $V_3\bar{x}_3 =$ Equation 3.41.
4. When $c > h$. $V_1 =$ Equation 3.34, $V_1\bar{x}_1 =$ Equation 3.35, $V_2 =$ Equation 3.38, $V_2\bar{x}_2 =$ Equation 3.39, $V_3 =$ Equation 3.42, and $V_3\bar{x}_3 =$ Equation 3.43.

Concrete force formulas for a square column are obtained by letting $b = d$ in the preceding formulas.

3.3 Bar forces

Bar force is a function of the point of zero strain and the limiting values of concrete and steel strain. In Figure 3.4 the compressive steel strain, e_s, is a function

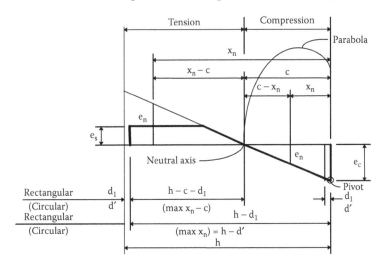

$$e_s = f_y/29{,}000 \text{ when } f_y \text{ is expressed in ksi}$$
$$= f_y/200{,}000 \text{ when } f_y \text{ is expressed in MPa}$$

$$e_c = 0.003 \text{ ACI method}$$
$$= 0.0035 \text{ Canadian method}$$

Figure 3.4 Steel bar strain diagram.
Source: Jarquio, Ramon V., *Analytical Method in Reinforced Concrete*. Universal Publishers. Boca Raton, FL. p. 59. Reprinted with permission.

of the useable concrete strain, e_c, in which the steel strain varies linearly from zero at the neutral axis to a maximum value of the concrete strain, $e_c = 0.003$.

Only the maximum value of $e_s = f_y/29,000$ (or $f_y/200,000$ in SI units) is used for ultimate strength calculations since 0.003 is greater than the steel yield strain, e_s.

$$\text{Tension or compression} = A_n f_{yn} \tag{3.46}$$

in which

$$A_n = \text{area of steel reinforcement}$$

$$A_n = \text{area of steel reinforcement}$$

$$f_{yn} = \text{steel stress} \leq f_y \tag{3.47}$$

$$f_{yn} = 29000\ e_n - f_{cn} \quad \text{(ksi)}$$

$$= 200000\ e_n - f_{cn}\ \text{(MPa)} \tag{3.48}$$

When $c > x_n$, the steel compressive strain is given by

$$e_n = (e_c/c)(c - x_n) \tag{3.49}$$

When $c < x_n$, the tensile strain is given by

$$e_n = (e_c/c)(x_n - c) \leq e_s \tag{3.50}$$

in which f_y = the yield stress of steel, c = the distance from the concrete compressive edge where strain and stress is zero, d' = the distance to the center of the first bar from the concrete edge, and x_n = the distance from the concrete edge to any steel bar.

From Figure 3.5,

$$f_{cn} = (0.75\ f_c'/c^2)(c - x_n)(c + 3x_n) \tag{3.51}$$

$$\text{Moment @ centerline} = A_n f_{yn}\ \{(h/2) - x_n\} \tag{3.52}$$

x_n values are determined as follows:

$$x_n = R - (R - d') \cos(n\ \alpha - \theta) \tag{3.53}$$

in which N = the total number of bars and n = the number of bar positions from 1 to N.

$$\alpha_1 = 2\pi/N \tag{3.54}$$

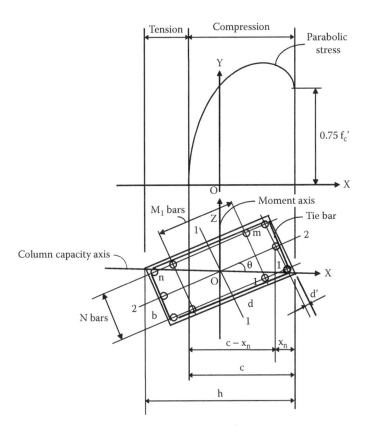

Figure 3.5 Steel reinforcing bars layout.

θ = the column capacity axis of the circular section. The envelope of values of θ is

$$0 < \theta \le \alpha_1 \tag{3.55}$$

When $\theta = \alpha_1$, the column capacity axis coincides with the center of a bar in the circular column section. As the column capacity axis is moved between any two bars, the moment capacity changes slightly. For practical purposes, the column capacity axis may be assumed to pass at the center of any bar and the center of the circular column section. Bar forces for the rectangular column section are calculated in the same manner as the circular column.

Values for x_n are calculated from Figure 3.5 as follows: For the first half of the bars along the width of the section,

$$x_n = (n - 1)[(b - 2d')/((N/2) - 1)]\sin \theta + d_1 \tag{3.56}$$

For the second half of the bars along the width of the section,

$$x_n = [(n - N/2) - 1][(b - 2d')/((N/2) - 1)]\sin \theta + (d - 2d')\cos \theta + d_1 \tag{3.57}$$

in which N = the number of bars along the width.

For half of the bars along the depth of the section,

$$x_n = m[(d - 2d')/((M_1/2) + 1)]\cos \theta + d_1 \tag{3.58}$$

For the other half of the bars along the depth of the section,

$$x_n = h - d_1 - [m - (M_1/2)][(d - 2d')/((M_1/2) + 1)] \cos \theta \tag{3.59}$$

in which n and m designate the nth and mth bars along width and depth, respectively, and M_1 = the number of bars along the depth.

$$\text{Total number of bars} = N + M_1 \tag{3.60}$$

$$d_1 = \sqrt{2}\,(d')\cos(45° - \theta) \tag{3.61}$$

From the preceding expressions, concrete and bar forces can be tabulated and combined to obtain the column capacity as measured by P, the axial capacity, and M, the moment capacity, and plotted to draw the column capacity curve. Excel will perform these calculations efficiently after the input data is entered. For a square column set, $b = d$ in these equations or identical numerical values are assigned to b and d in the calculations for ultimate strength.

For design purposes, the ultimate strength capacity should be calculated along the diagonal of the rectangular section. For this case, the following relationships are applicable:

$$\theta = \alpha = \arctan(b/d) \tag{3.62}$$

$$h = (b^2 + d^2)^{1/2} \tag{3.63}$$

$$z_0 = 0 \tag{3.64}$$

3.4 Capacity curves

From the preceding equations, we can calculate the ultimate strength capacities of the column section using the relationships

$$P = P_c + P_s \tag{3.65}$$

$$M = M_c + M_s \tag{3.66}$$

in which P and M are the axial and moment capacities, respectively, of the column, P_c and M_c are the concrete forces, and P_s and M_s are the steel bar forces.

Example 3.1

Assume a circular column 24 in. in diameter reinforced with 16 #7 (22 mm) bars, with $f_c' = 5$ ksi, $f_y = 60$ ksi, and $e_c = 0.003$. Plot the ultimate strength capacity of this column when the capacity axis is along a diameter and through the center of a bar in the column section.

> **Solution:** Substitute the given values in the Excel spreadsheets and the equations for circular column sections. The concrete forces are calculated using Equations 3.04 to 3.17 and plotted in Figure 3.6. Next, the steel bar forces are determined from Equations 3.46 to 3.55 and plotted in Figure 3.7. Combine the results to obtain the capacity curve of the circular column in Figure 3.8.

Example 3.2

A rectangular column 23 in. (58.42 cm) × 17 in. (43.18 cm) with 14 #7 (22-mm) bars and d' = 2.50 in. (6.35 cm), N = 10 bars, Ml = 4 bars, fc' = 5 ksi (34.5 MPa), f_y = 60 ksi (414 MPa), and e_c = 0.003. Plot the capacity curve along the diagonal of the rectangular section when $\theta = \alpha = \arctan(17/23) = 0.635$ radians.

> **Solution:** Substitute the given values in the Excel spreadsheets and the equations for rectangular column sections. The concrete forces are calculated using Equations 3.32 to 3.45 and plotted in

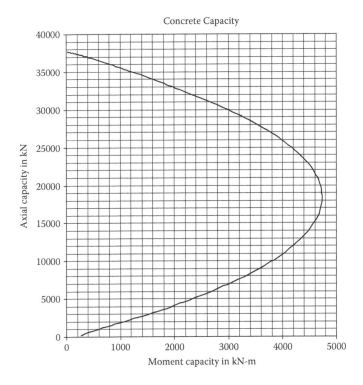

Figure 3.6 Concrete capacity for circular column.

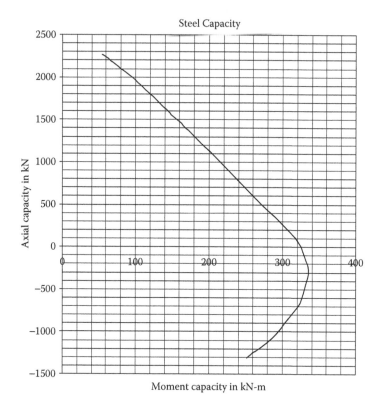

Figure 3.7 Steel capacity for a circular column.

Figure 3.9. Next, the steel bar forces are determined from Equations 3.56 to 3.61 and plotted in Figure 3.10. Combine the results to obtain the capacity curve of the circular column in Figure 3.11.

The capacity curve in Figure 3.11 represents the component of the resultant moment capacity along the diagonal of the rectangular column. To use this capacity curve for plotting the external loads for determining whether the loads are within the envelope of the capacity curve, it is necessary to obtain the component of the resultant external moment along the diagonal. This component is the one to be plotted for comparison to the capacity curve. The calculation of the resultant moment capacity is discussed in Section 3.5, which illustrates the properties of the capacity axis as well as the calculation of the moment capacity perpendicular to the capacity axis. The capacity curve indicates the interplay of the axial capacity with the moment capacity as the position of the neutral axis is varied. Each position of the neutral axis represents a specific column capacity. The capacity curve indicates that as the axial load is diminished, the moment capacity increases. The bending moment is at its greatest value when the compressive depth c is at balanced condition, i.e., when the steel tensile stress and concrete stress are developed

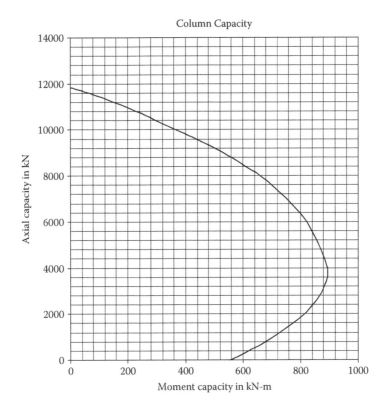

Figure 3.8 Circular column capacity curve.

simultaneously. When the axial load is zero, the beam condition develops. A pure axial loading condition is difficult to attain because there is always an eccentricity present in the column section. Since a column section is mostly under biaxial loading conditions, the eccentricity is given by $e = M/P$. This eccentricity represents the internal conditions of the column.

The eccentricity of the external loads includes the buckling of the column for a specific length, out-of-plumb measurements, and end moments at supports. These add additional external bending moments. The total external bending moment is then compared to the capacity curve to determine the column section's ability to resist the external load. The development of the capacity curve does not require the application of the standard interaction formula for biaxial bending or the application of Euler's column formula, which is more suitable for determining the external load.

The determination of external loads differs for every location and applicable code of practice. This is where the expertise of the structural engineer is required because of the many unpredictable loading conditions that can occur. Meanwhile, the determination of internal capacity is only the function of the geometry of the section and stress–strain characteristics of the structural material. In order words, it is predictable using only basic mathematics.

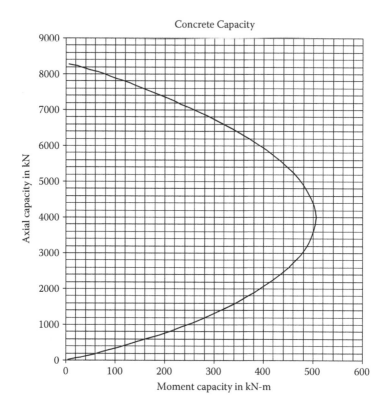

Figure 3.9 Concrete capacity for a rectangular column.

3.4.1 Key points

The customary key points are shown in Figure 3.12 and described as follows:

- Key point Q is determined when the moment of the concrete forces is equated to the moment of the bar forces. The maximum value of c at this point develops the theoretical maximum axial load capacity.
- Key point T is the point where the maximum moment capacity of the column section is developed. This point coincides with the value at the balanced condition of loading designated as key point U for the CRSI stress–strain diagram. For other stress–strain assumptions, such as those recommended by the author in his first book, *The Analytical Method in Reinforced Concrete*, key point T can occur for rectangular the column section.
- Key point V is determined when the concrete forces are equated to the bar forces. The minimum value of c at this point will develop the moment capacity of the column section as a doubly reinforced concrete beam. The rest of the key points, R, S, and U, are simply determined as the sums of the concrete and bar forces at specific values of c.

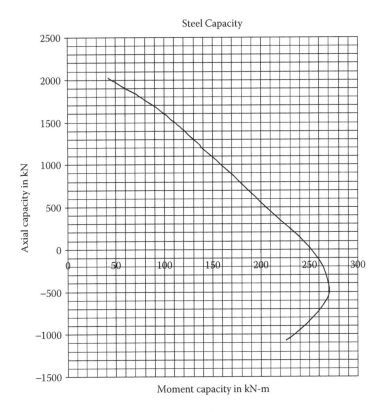

Figure 3.10 Steel capacity for a rectangular column.

- Key point R is the point where $c = h$ (full column section in compression). This is the point of minimum eccentricity to use in a design in which the corresponding axial capacity of the column is the maximum allowable for the column section under consideration. The ACI method determines the maximum axial load by an empirical formula applicable to tied and spirally reinforced short columns.
- Key point S is the point in the column section where the tension of the bar farthest from the concrete edge is zero. From this point the bar strain increases linearly to the maximum at the balanced condition.
- Key point U is the commonly designated balanced condition of loading. At this point, the maximum tensile steel strain, e_s, and the useable concrete compressive strain, e_c, occur simultaneously. Contrary to popular belief, this is not always the point of maximum moment resistance of a column section, depending on the shape of the stress–strain diagram at the tensile portion of the section. Hence, key point T is added to the list of key points in the column capacity curve. There may be column conditions in which these two points coincide or interchange their relative positions in the column capacity curve.

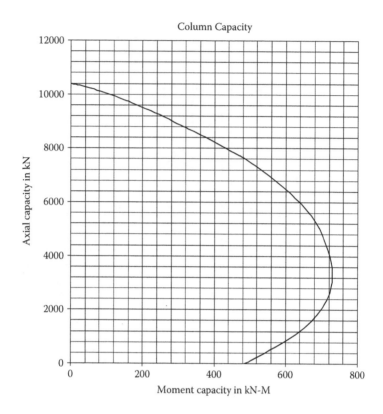

Figure 3.11 Rectangular column capacity curve.

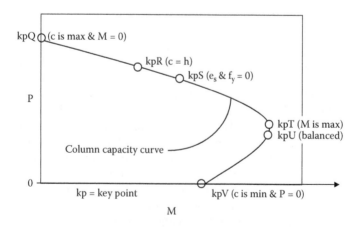

Figure 3.12 Schematic locations of key points.

Table 3.1 Key Point Capacities of Circular Column Example (CRSI Stress–Strain Method)

Key Point	c	Ms	Ps	Mc	Pc	M	P	e=M / P
Q	43.536	486	510	-486	2149	0	2658	
R	24.000	1291	357	2545	1803	3836	2160	1.776
S	21.500	1479	318	3635	1627	5114	1946	2.629
T	12.297	2882	1	5046	831	7928	832	9.529
U	12.724	2803	24	5110	869	7913	893	8.857
V	5.814	2228	-294	2685	295	4913	0	Very Large

For Example 3.1, the numerical values at key points for the CRSI method are shown in Table 3.1. The values of the key points for the author's method illustrated in his first book are shown in Table 3.2. For Example 3.2, the numerical values at key points for the CRSI method are shown in Table 3.3, and the author's are shown in Table 3.4.

Comparing the two methods for the circular column, we note that they have common values at key points, Q, R, and S. They only differ in values at key points T, U, and V. The author's method gives a lower value of moment capacity at key point T (7912 versus 7928) than the CRSI method. At key point V the author's method also gives a lower moment capacity (3963 versus 4913).

The difference occurs because of the different stress–strain conditions assumed for calculations when the position of the neutral axis defined by the variable c is less than at the balanced condition, resulting in the tension side of the column section exceeding the yield stress and strain of the reinforcing bars. The author's method limits the stress–strain condition of the bar farthest from the neutral axis to f_y, while the CRSI method allows the

Table 3.2 Key Point Capacities of Circular Column (Author's Stress–Strain Method)

Key Point	c	Ms	Ps	Mc	Pc	M	P	e = M / P
Q	43.536	486	510	-486	2149	0	2658	
R	24.000	1291	357	2545	1803	3836	2160	1.776
S	21.500	1479	318	3635	1627	5114	1946	2.629
T	12.725	2803	24	5110	869	7912	893	8.856
U	12.724	2803	24	5110	869	7913	893	8.857
V	4.647	1896	-213	2066	213	3963	0	Very Large

Table 3.3 Key Point Capacities of Rectangular Column Example (CRSI Stress–Strain Method)

Key Point	c	Ms	Ps	Mc	Pc	M	P	e=M / P
Q	52.340	374	455	-374	1880	0	2334	.000
R	28.601	1045	321	1898	1634	2943	1955	1.505
S	25.105	1207	277	2947	1460	4154	1737	2.391
T	14.858	2201	10	4257	688	6458	698	9.253
U	14.858	2201	10	4257	688	6458	698	9.253
V	8.570	2001	-240	2281	240	4282	0	very large

stress and strain to be pivoted at the maximum useable concrete strain equal to 0.003. For design purposes, the author's method for determining circular column capacities is preferable and conservative because of lower moment values obtained than that by using the CRSI method.

For Example 3.2, the methods differ in values at key points T, U, and V. The author's method gives a higher value of moment capacity at key point T (6490 versus 6458) than the CRSI method, although at key point U both have the same value. At key point V the author's method also gives a lower moment capacity (3371 versus 4282). The difference occurs for the reasons just described. The author's method limits the stress–strain condition of the bar farthest from the neutral axis to f_y, while the CRSI method allows the stress and strain to be pivoted at the maximum useable concrete strain equal to 0.003. For design purposes the CRSI method illustrated in this book (not the current method by CRSI) is preferable than the one shown by the author in his first book, *Analytical Method in Reinforced Concrete*.

Table 3.4 Key Point Capacities of Rectangular Column Example (Author's Stress–Strain Method)

Key Point	c	Ms	Ps	Mc	Pc	M	P	e = M / P
Q	52.340	374	455	-374	1880	0	2334	.000
R	28.601	1045	321	1898	1634	2943	1955	1.505
S	25.105	1207	277	2947	1460	4154	1737	2.391
T	15.842	2103	44	4387	766	6490	811	8.005
U	14.858	2201	10	4257	688	6458	698	9.253
V	7.278	1601	-173	1770	173	3371	0	very large

3.5 Column capacity axis

The column capacity axis is the line perpendicular to the moment axis. For a circular column section, the load axis (the line passing through the point of external load and the center of the circular column) can be considered to fall between any two steel bars. The point of external load application is located with the reference axis by $\theta_u = \arctan(M_2/M_1)$, in which M_2 and M_1 are the external bending moments around the X-axis and the Z-axis, respectively. The column capacity axis can be positioned from $\theta = 0$ to $\theta = \alpha_1$.

The column capacity axis and load axis are not always coincident, because of the nonsymmetry of the bar forces. The column capacity axis may be assumed to pass through the center of any steel bar because the moment variations are small. The actual column capacity axis for equilibrium of external and internal forces may be determined by using the lever arm of any steel bar as given by the expression

$$z = (R - d') \sin(n\,\alpha_1 - \theta) \tag{3.67}$$

For a rectangular column section, the column capacity axis establishes the equilibrium of external and internal forces. For the equilibrium condition, the point of application of external forces must coincide with the centroid of the internal forces; that is, the external and internal forces must be collinear. Figure 3.13 shows how the external load has to be applied to a column section to ensure this equilibrium.

$$e = e_u \cos(\theta - \theta_u) \tag{3.68}$$

$$P = P_u \tag{3.69}$$

c.g. = Centroid of internal forces shown here coincident with the position of the external load for equilibrium

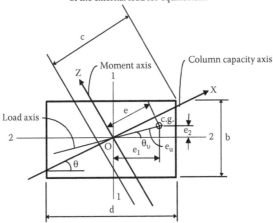

Figure 3.13 Column capacity axis and load application.
Source: Jarquio, Ramon V. *Analytical Method in Reinforced Concrete.* Universal Publishers, Boca Raton, FL. p. 83. Reprinted with Permission.

in which e = the eccentricity of the ultimate capacity of the column section (M/P), e_u = the eccentricity of the given loading condition (M_u/P_u), θ_u = the angle of eccentricity of the given loading condition (M_2/M_1), and θ = the angle of the column capacity axis with the horizontal axis.

Equation 3.68 indicates that the resultant external load falls below the column capacity axis for a particular value of c in the rectangular column section. In other words, we cannot use the column capacity along the load axis since the resultant external load will not coincide with the center of gravity of internal forces of a column section defined by the value of c.

The author recommends that, for design purposes, we assume the diagonal as the reference axis for column capacity since the moment capacity increases toward the uniaxial position of the position of the column capacity at $\theta = 0$. This assumption allows the preparation of column capacity curves or tables for any column section. It also ensures that the values being used are within the envelope of the column capacity curve. Once the plot of the capacity curve is obtained, the external load, $P_u e_u$, should be resolved along the diagonal such that the external moment given by $P_u e_u \cos(\theta - \theta_u)$ can be plotted on the capacity curve, if the supplied section is adequate to support the external loads within the envelope of the capacity curve.

The following equations should be used to check for equilibrium conditions for a column section subjected to an external biaxial load equal to $P_u e_u$:

$$\sum Vz = P_u e_u \sin(\theta - \theta_u) \tag{3.70}$$

$$\bar{z} = \sum Vz \Big/ \sum V = e_u \sin(\theta - \theta_u) \tag{3.71}$$

$$\sum V = P_u \tag{3.72}$$

For concrete forces (see Figure 3.3),

$$V_2 z_2 = -0.50(\cot\theta - \tan\theta)V_2\,\bar{x}_2 - [z_0 - 0.25\,h\,(\cot\theta - \tan\theta)]V_2 \tag{3.73}$$

$$V_1 z_1 = \tan\theta\,(V_1\,\bar{x}_1) \quad \text{when} \quad \theta < [(\pi/2) - \alpha] \tag{3.74}$$

$$V_1 z_1 = -\cot\theta\,(V_1\,\bar{x}_1) \quad \text{when} \quad \theta > [(\pi/2) - \alpha] \tag{3.75}$$

$$V_3 z_3 = -0.50(\cot\theta - \tan\theta)\,V_3\,\bar{x}_3 - [0.25\,h(\cot\theta - \tan\theta) - z_0]V_3 \tag{3.76}$$

For bar forces lever arms (see Figure 3.5):

- M bars:

$$+z = (x_m - d_1)\tan\theta < (d - 2d')\sin\theta \tag{3.77}$$

$$-z = [h - x_m - d_1]\tan\theta < (d - 2d')\sin\theta \tag{3.78}$$

- N bars:

$$+z = (h - x_n - d_1)\cot\theta < (d - 2d')\sin\theta \tag{3.79}$$

$$-z = (x_n - d_1)\cot\theta < (d - 2d')\sin\theta \tag{3.80}$$

Use the initial value of θ at the diagonal and combine moments of the concrete and bar forces to solve for the initial value of $\overline{z}$. Compare this value with $e_u \sin(\theta - \theta_u)$. Repeat until these two values are almost equal to obtain the correct column capacity axis for an external load of $P_u e_u$. This will complete the solution for the true column capacity axis to use for a column section subjected to a specific loading condition.

To ensure safety, the internal capacity should always be greater than the external load so that all external load combinations such as those in bridge design loading are inside the envelope of the column capacity curve.

3.5.1 Variation of moment capacity

As the column capacity axis varies from zero to $\pi/2$, the moment capacity also varies accordingly. Figure 3.14 shows the variation of moment capacity at key point U of the column capacity curve. Figure 3.15 is the variation of moment capacity at key point S as θ is varied from zero to $\pi/2$. It is obvious from this demonstration of moment capacity that the standard column interaction formula for biaxial bending, which gives only one set of column capacity values, is no longer warranted in structural analysis.

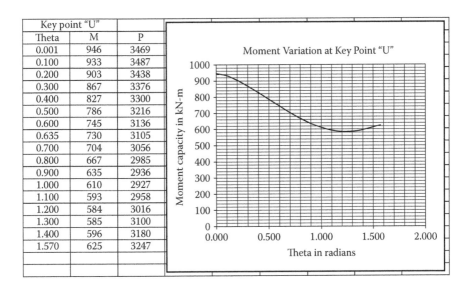

Key point "U"		
Theta	M	P
0.001	946	3469
0.100	933	3487
0.200	903	3438
0.300	867	3376
0.400	827	3300
0.500	786	3216
0.600	745	3136
0.635	730	3105
0.700	704	3056
0.800	667	2985
0.900	635	2936
1.000	610	2927
1.100	593	2958
1.200	584	3016
1.300	585	3100
1.400	596	3180
1.570	625	3247

Figure 3.14 Moment variation at key point U.

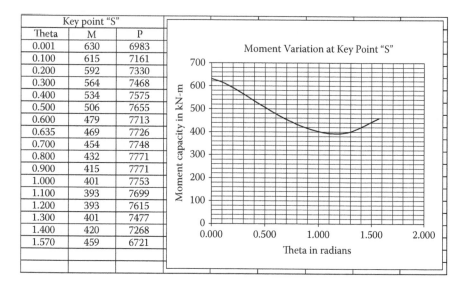

Key point "S"		
Theta	M	P
0.001	630	6983
0.100	615	7161
0.200	592	7330
0.300	564	7468
0.400	534	7575
0.500	506	7655
0.600	479	7713
0.635	469	7726
0.700	454	7748
0.800	432	7771
0.900	415	7771
1.000	401	7753
1.100	393	7699
1.200	393	7615
1.300	401	7477
1.400	420	7268
1.570	459	6721

Figure 3.15 Moment variation at key point S.

It is also clear that for biaxial bending conditions, development of the uniaxial capacity at $\theta = 0$ and $\theta = \pi/2$ is not possible. This alone makes the application of the standard interaction formula for biaxial bending not only incorrect, but also a violation of the basic principle of structural mechanics that requires consideration of the geometric properties of the section.

3.5.2　Limitations of the standard interaction formula

The significant limitations of the standard interaction formula can be shown from the results of the analytical method. The sum of the ratios of the external load to the internal capacity of a section in orthogonal axes 2-2 and 1-1 can be added up to test for unity as a maximum value. From the tabulated values, it can be easily concluded that application of the standard interaction formula will result in the underutilization of a structural member.

Figure 3.16 shows the capacity curve for Example 3.2 (584 × 432 mm with 14 to 22 mm diameter reinforcing steel bars). Concrete stress $f_c' = 34.5$ MPa and the useable concrete strain = 0.003. The yield stress of the reinforcing steel bar = 414 MPa. The column capacity axis is at the diagonal of the rectangular column. The test is applied at key point U of the capacity curve. The capacity curve of the column is taken from the Excel worksheet in which the uniaxial capacities of this column are $M_1 = 946$ kN-m, $M_2 = 625$ kN-m, $M = 730$ kN-m, and $Mz = 235$ kN-m. These values will be used in the interaction formula to demonstrate the limitations of its effectiveness in predicting the capacity of a column section.

Table 3.5 shows the application of the interaction formula using the results of the analytical method for Example 3.2. Assume that the resultant

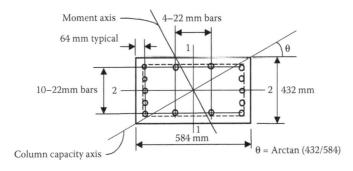

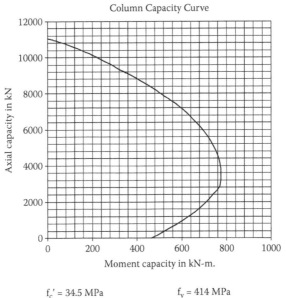

Column Capacity Curve

$f_c' = 34.5$ MPa $f_y = 414$ MPa

Figure 3.16 Column capacity curve.

internal moment capacity is equal to the external load. The resultant bending moment capacity is obtained as $M_R = \{(730)^2 + (235)^2\}^{1/2} = 767$ kN-m.

The load axis and location of the resultant bending moment from the horizontal is given by the expression

$$\theta_u = \alpha - \beta \qquad (3.81)$$

in which $\beta = \arctan (M_z/M) = \arctan (235/730) = 0.3114$ radians and $\alpha = \arctan (b/d) = \arctan (17/23) = 0.6365$ radians.

Substitute the values in Equation 3.81 to obtain $\theta_u = 0.6365 - 0.3114 = 0.3251$ radians. The result is resolved along the 1 and 2 axes as

$$M_1' = M_R' \cos \theta_u \quad \text{and} \quad M_2' = M_R' \sin \theta_u \qquad (3.82)$$

Table 3.5 Application of the Interaction Formula to Example Rectangular Column

P =	3105	MR =	767	Beta =	0.3114	PMAX =	10382		
M =	730	M1 =	946	Alpha =	0.6365				
Mz =	235	M2 =	625	Theta U =	0.3251				
M1'	M2'	MR'	R1	R2	R1 + R2	R3	R1 + R2 + R3	R	%
727	245	767	0.768	0.392	1.160	0.299	1.459	1.000	100
711	240	750	0.751	0.383	1.135	0.299	1.434	0.978	98
687	232	725	0.726	0.370	1.097	0.299	1.396	0.945	95
663	224	700	0.701	0.358	1.059	0.299	1.358	0.913	91
640	216	675	0.676	0.345	1.021	0.299	1.320	0.880	88
616	208	650	0.651	0.332	0.983	0.299	1.282	0.848	85
592	200	625	0.626	0.319	0.945	0.299	1.245	0.815	81
569	192	600	0.601	0.307	0.908	0.299	1.207	0.782	78
545	184	575	0.576	0.294	0.870	0.299	1.169	0.750	75
521	176	550	0.551	0.281	0.832	0.299	1.131	0.717	72
498	168	525	0.526	0.268	0.794	0.299	1.093	0.685	68
474	160	500	0.501	0.255	0.756	0.299	1.055	0.652	65
450	152	475	0.476	0.243	0.719	0.299	1.018	0.619	62
426	**144**	**450**	**0.451**	**0.230**	**0.681**	**0.299**	**0.980**	**0.587**	**59**
403	136	425	0.426	0.217	0.643	0.299	0.942	0.554	55
379	128	400	0.401	0.204	0.605	0.299	0.904	0.522	52
355	120	375	0.376	0.192	0.567	0.299	0.866	0.489	49
332	112	350	0.351	0.179	0.529	0.299	0.829	0.456	46

The ratios of external to internal moments are $R_1 = M_1'/M_1$ and $R_2 = M_2'/M_2$. $R = M_R'/M_R$, which is the fraction of the resultant bending moment utilized in the interaction formula. These relationships and above numerical values are used with the standard interaction formula for biaxial bending to construct Table 3.5, in which the usefulness of the standard interaction formula is below 59% of its potential ultimate strength capacity. From these results, we now know that the accuracy of the standard interaction formula is substandard and therefore is no longer warranted and should be discarded.

Notations

b: width of a rectangular concrete section
c: depth of a concrete section in compression
d: depth of a rectangular concrete section
d': concrete cover from the edge to center of any steel bar
d_1: distance from the concrete edge to the first steel bar
e: eccentricity $= M/P$
e_c: useable concrete strain
e_s: steel yield strain
e_n: compressive (or tensile) steel strain at the nth bar location
e_u: eccentricity of the external load (M_u/P_u)
f_c: compressive stress in concrete
f_c': specified ultimate compressive strength of concrete
f_y: specified yield stress of a steel bar
f_{yn}: steel stress $\leq f_y$
h: overall thickness of a member

x_m: location of a steel bar from a reference axis

x_n: location of a steel bar from the concrete compressive edge

$\bar{z}$: distance from the X-axis of the centroid of internal forces

M_1: external moment around axis 1-1

M_2: external moment around axis 2-2

M: internal ultimate moment capacity

M_l: number of bars along the depth of a section

M_u: external resultant moment

N: number of bars along the width of a section

P: internal ultimate axial capacity

P_u: external axial load

R: radius of a circular column

θ: inclination of the column capacity axis with the horizontal axis

θ_u: arctan M_2/M_1 (inclination of the resultant external forces about the horizontal axis)

α_1: central angle subtended by one bar spacing in a circular column

α: arctan (b/d) = the inclination of the diagonal of a rectangular column with the horizontal axis

ACI: American Concrete Institute

CRSI: Concrete Reinforcing Steel Institute

Note: All other alphabets and symbols used in mathematical derivations are defined in the context of their use.

chapter four

Concrete-filled tube columns

4.1 Introduction

The current method of calculating the ultimate strength of concrete-filled tube (CFT) columns employs the column interaction formula for steel and reinforced concrete sections subjected to biaxial bending. In contrast, the analytical method illustrated in this book will eliminate the need to use the column interaction formula by using the column capacity axis not only as the reference for equilibrium of internal and external forces, but also to determine the capacities of the column section at every position of this axis.

This analysis involves calculating the concrete and steel forces separately and then combining these to determine the ultimate strength of CFT columns. The equations for the steel forces were presented at the ISEC-02 conference in Rome. The assumptions used in the ultimate strength of reinforced concrete columns are also applicable in analyzing the ultimate strength of CFT column sections. Equations for determining the centroid of internal forces and factors for external loads are also applicable in this case. At ultimate conditions of stress and strain, the concrete strain, usually assumed equal to 0.003 by the American Concrete Institute (ACI) method and 0.0035 by Canadian practice, is the pivot point for determining the steel stress and strain. The concrete and steel elements in a CFT column section undergo common deformation when resisting external loads.

This deformation is assumed linear with respect to the neutral axis, whose location is the concrete compressive depth c. As c is varied from the concrete edge it will generate a steel stress–strain diagram consisting of a triangular and rectangular shape. This steel stress–strain diagram will define the steel forces to be added to concrete forces to obtain the ultimate strength of a CFT column section. The compressive and tensile steel stress is limited to the yield strength f_y of the material.

The derived equations for circular and rectangular CFT columns are programmed using Microsoft Excel 97 to generate capacity curves for these columns. From the numerical data of a rectangular CFT column, the accuracy of the current standard interaction formula for biaxial bending can be shown to be substandard (see Table 4.1 for details).

Table 4.1 Limitations of the Standard Interaction Formula for Biaxial Bending for CFT Rectangular Column

M₁'	M₂'	MR'	R₁	R₂	R₁ + R₂	R₃	R₁ + R₂ + R₃	R	%
852	609	1047	0.693	0.613	1.306	0.134	1.440	1.00	100
834	596	1025	0.679	0.600	1.279	0.134	1.412	0.98	98
814	581	1000	0.662	0.585	1.247	0.134	1.381	0.96	96
793	567	975	0.646	0.571	1.216	0.134	1.350	0.93	93
773	552	950	0.629	0.556	1.185	0.134	1.319	0.91	91
753	538	925	0.612	0.541	1.154	0.134	1.288	0.88	88
732	523	900	0.596	0.527	1.123	0.134	1.256	0.86	86
712	509	875	0.579	0.512	1.092	0.134	1.225	0.84	84
692	494	850	0.563	0.498	1.060	0.134	1.194	0.81	81
671	480	825	0.546	0.483	1.029	0.134	1.163	0.79	79
651	465	800	0.530	0.468	0.998	0.134	1.132	0.76	76
631	450	775	0.513	0.454	0.967	0.134	1.100	0.74	74
610	436	750	0.497	0.439	0.936	0.134	1.069	0.72	72
590	421	725	0.480	0.424	0.904	0.134	1.038	0.69	69
570	407	700	0.463	0.410	0.873	0.134	1.007	0.67	67
549	**392**	**675**	**0.447**	**0.395**	**0.842**	**0.134**	**0.976**	**0.64**	**64**
529	378	650	0.430	0.380	0.811	0.134	0.945	0.62	62
509	363	625	0.414	0.366	0.780	0.134	0.913	0.60	60
488	349	600	0.397	0.351	0.748	0.134	0.882	0.57	57
468	334	575	0.381	0.337	0.717	0.134	0.851	0.55	55
448	320	550	0.364	0.322	0.686	0.134	0.820	0.53	53
427	305	525	0.348	0.307	0.655	0.134	0.789	0.50	50
407	291	500	0.331	0.293	0.624	0.134	0.757	0.48	48
387	276	475	0.315	0.278	0.593	0.134	0.726	0.45	45
366	262	450	0.298	0.263	0.561	0.134	0.695	0.43	43
346	247	425	0.281	0.249	0.530	0.134	0.664	0.41	41
325	232	400	0.265	0.234	0.499	0.134	0.633	0.38	38

4.2   Derivation

The governing equations for determining the ultimate strength of a column section are

$$P = P_c + P_s \tag{4.1}$$

$$M = M_c + M_s \tag{4.2}$$

The subscripts c and s for P and M denote concrete and steel capacities, respectively. The concrete forces were derived in Chapter 3. Hence, only the equations for the steel forces need to be derived.

4.2.1   Steel forces for circular sections

In Figure 4.1, from the equilibrium conditions of external and internal forces acting on the column section, the following relationships become apparent:

$$\sum F = 0 : P = C_c + C_s - T_s \tag{4.3}$$

$$M = C_c x_c + C_s x_s + T_s x_t \tag{4.4}$$

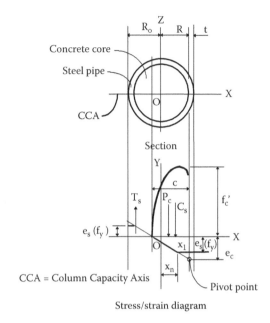

Figure 4.1 Circular CFT column section.
Source: Jarquio, Ramon V. *Analytical Method in Reinforced Concrete.* Universal Publishers, Boca Raton, FL. p. 95. Reprinted with Permission.

The steel forces are represented by C_s, T_s, $C_s x_s$, and $T_s x_t$. To determine the steel tensile and compressive forces, designate three main sections such as V_1, V_2, and V_3. V_1 is the compressive steel stress volume due to the triangular stress value, V_2 is the compressive steel stress volume due to the uniform stress value, f_y, and V_3 is the tensile steel stress volume due to a varying stress value, f_y. These volumes vary as a function of the value of c.

Plot the stress–strain diagram such that the concrete reaches its assigned value of strain and the steel reaches the maximum yield stress, f_y. Write the governing equations for the tensile and compressive stress–strain lines to calculate the steel forces. This will require six limiting envelopes for c to determine the values of the steel forces.

The equations for the outer circular section are derived first, followed by the inner circular section. From analytic geometry, the equation of a circle is as follows:

$$x^2 + z^2 = R_o^2 \qquad (4.5)$$

The equation for the triangular stress is given by

$$y = k \{x + (c - R)\} \qquad (4.6)$$

in which

$$k_1 = 29000e_c/c \tag{4.7}$$

The equation for the uniform stress is given by

$$y = k_2 = f_y \tag{4.8}$$

The intersection of the uniform stress with triangular stress is given by

$$x_1 = (1/e_c)c(e_c - e_s) \tag{4.9}$$

$$x_n = R - x_1 \tag{4.10}$$

Evaluate the integrals of the stress volumes within their respective limits defined by the limiting ranges of c from the compressive concrete edge as it passes through the geometric limits of these three volumes and the strain diagram limits for the uniform and triangular shapes. Figure 4.2 shows the stress–strain variables as functions of c. This process should yield 35 equations for the outer circular section. Repeat the procedure to obtain another 35 equations for the inner circular section. The difference between the outer and inner expressions is the value of the steel forces of the CFT column. Add the concrete forces to this to obtain the ultimate strength of the CFT column for every value of c on a column section. Plot the values of P and M to yield the ultimate strength capacity curve of the CFT column section.

The following are the derived equations for the outer circular section for every envelope of values of c on the CFT column section.

Case 1: $0 < c < c_0$.

$$V_2 = k_2 \left\{ 1.5708R_o^2 - \left[x_n \left(R_o^2 - x_n^2 \right)^{1/2} + R_o^2 \arcsin(x_n/R_o) \right] \right\} \tag{4.11}$$

$$V_1 = (k_1/3) \left\{ 2\left(\left[R_o^2 - (R-c)^2 \right]^{3/2} - \left(R_o^2 - x_n^2 \right)^{3/2} \right) + 3(c-R)\left(\left[x_n \left(R_o^2 - x_n^2 \right)^{1/2} \right. \right. \right.$$
$$\left. \left. + R_o^2 \arcsin \left(x_n/R_o \right) \right] - (R-c)\left[R_o^2 - (R-c)^2 \right]^{1/2} + R_o^2 \arcsin[(R-c)/R_o] \right\} \tag{4.12}$$

$$V_{3a} = -[k_1/3] \left\{ -2\left[R_o^2 - (R-c)^2 \right]^{3/2} - 3(c-R)\left[x_p \left[R_o^2 - x_p^2 \right]^{1/2} + R_o^2 \arcsin(x_p/R_o) \right] \right\} \tag{4.13}$$

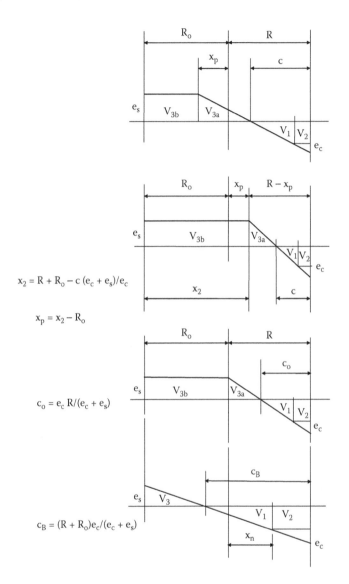

Figure 4.2 Stress–strain variables.

$$V_{3b} = k_2 \left\{ x_p \left(R_o - x_p^2 \right)^{1/2} + R_o^2 [1.5708 + \arcsin(x_p/R_o)] \right\}$$ (4.14)

$$C_s = V_1 + V_2$$ (4.15)

$$T_s = V_{3a} + V_{3b}$$ (4.16)

$$V_1 x_1 = (k_1/12) \left\{ -6\, x_n \left[R_o^2 - x_n^2 \right]^{3/2} + 3R_o^2 \left[x_n \left(R_o^2 - x_n^2 \right)^{1/2} \right. \right.$$

$$+ R_o^2 \, \arcsin\, (x_n/R_o) \bigg] + 6(R - c)\left(\left[R_o^2 - (R - c)^2 \right]^{3/2} \right.$$

$$- 3R_o^2 [R - c]\left[R_o^2 - (R - c)^2 \right]^{1/2} + R_o^2 \, \arcsin\, [(R - c)/R_o] \Big)$$

$$\left. - 8\,(c - R)\left(\left[R_o^2 - x_n^2 \right]^{3/2} - \left[R_o^2 - (R - c)^2 \right]^{3/2} \right) \right\} \tag{4.17}$$

$$V_2 x_2 = (2k_2/3)\left(R_o^2 - x_n^2 \right)^{3/2} \tag{4.18}$$

$$V_{3a} x_{3a} = (k_1/12)\left\{ \left(R_o^2 - x_p^2 \right)^{3/2} (6x_p + 8[c - R]) - \left[R_o^2 - (c - R)^2 \right]^{3/2} (2c - 2R) \right.$$

$$+ 3R_o^2 \left(-x_p \left[R_o^2 - x_p^2 \right]^{1/2} + R_o^2 \, \arcsin(x_p/R_o) + (R - c)\left[R_o^2 - (R - c)^2 \right]^{1/2} \right.$$

$$\left. + R_o^2 \, \arcsin[(R - c)/R_o] \right) \bigg\} \tag{4.19}$$

$$V_{3b} x_{3b} = (2k_2/3)\left(R_o^2 - x_p^2 \right)^{3/2} \tag{4.20}$$

$$C_s x_s = V_1 x_1 + V_2 x_2 \tag{4.21}$$

$$T_s x_t = V_3 x_{3a} + V_{3b} x_{3b} \tag{4.22}$$

Case 2: $c_o < c < R$. $V_2 =$ Equation 4.11, $V_2 x_2 =$ Equation 4.18, and $V_{3b} x_{3b} =$ Equation 4.20.

$$V_1 = (k_1/3)\left\{ -2\left(\left[R_o^2 - x_n^2 \right]^{3/2} - \left[R_o^2 - (c - R)^{1/2} \right] + 3(c - R)\left(x_n \left[R_o^2 - x_n^2 \right]^{1/2} \right. \right. \right.$$

$$\left. + (c - R)\left[R_o^2 - (c - R)^2 \right]^{1/2} + R_o^2 \, \arcsin(x_n/R_o) + \arcsin[(c - R)/R_o] \right) \bigg\} \tag{4.23}$$

$$V_1 x_1 = (k_1/12)\left\{ 3R_o^2 x_n \left(R_o^2 - x_n^2 \right)^{1/2} + 3R_o^2 \,(c - R)\left[R_o^2 - (c - R) \right]^{1/2} \right.$$

$$+ 3R_o^4 \,(\arcsin(x_n/R_o) + \arcsin[(c - R)/R_o]) - 6x_n \left(R_o^2 - x_n^2 \right)^{3/2} -$$

$$6(c - R)\left[R_o^2 - (c - R)^2 \right]^{3/2} - 8(c - R)\left(R_o^2 - x_n^2 \right)^{3/2} +$$

$$\left. 8(c - R)\left[R_o^2 - (c - R)^2 \right]^{3/2} \right\} \tag{4.24}$$

$$V_{3a} = -(k_1/3) \left\{ -2\left[R_o^2 - (R-c)^2\right]^{3/2} + 3(c-R)\left([R-c]\left[R_o^2 - (R-c)^2\right]^{1/2}\right.\right.$$

$$\left. + R_o^2 \ \arcsin[(R-c)/R_o]\right) + 2\left[R_o^2 - x_p^2\right]^{3/2} + 3(c-R)\left(x_p\left[R_o^2 - x_p^2\right]^{1/2}\right.$$

$$\left.\left. + R_o^2 \ \arcsin[x_p/R_o]\right)\right\} \tag{4.25}$$

$$V_{3b} = k_2 \left\{ -x_p \left(R_o^2 - x_p^2\right)^{1/2} + R_o^2 \ [1.5708 - \arcsin(x_p/R_o)]\right\} \tag{4.26}$$

$$V_{3a}x_{3a} = (K_1/12) \left\{ \left(R_o^2 - x_p^2\right)^{3/2} (6x_p - 8[c-R]) + \left[R_o^2 - (R-c)^2\right]^{3/2} (2c - 2R)\right.$$

$$- 3R_o^2\left(x_p \ \left[R_o^2 - x_p^2\right]^{1/2} + R_o^2 \ \arcsin[x_p/R_o]\right) + (R-c)\left(\left[R_o^2 - (R-c)^2\right]^{1/2}\right.$$

$$\left.\left. + R_o^2 \ \arcsin[(R-c)/R_o]\right)\right\} \tag{4.27}$$

Case 3: $R < c < c_B$. V_2 = Equation 4.11, V_2x_2 = Equation 4.18, V_{3a} = Equation 4.25, V_{3ax3a} = Equation 4.27, V_{3b} = Equation 4.26, $V_{3b}x_{3b}$ = Equation 4.20, and

$$c_B = [e_c/(e_c + e_s)](R + R_o) \tag{4.28}$$

$$V_1 = (k_1/3) \left\{ -2 \left[\left(R_o^2 - x_n^2\right)^{3/2} - \left[R_o^2 - (c-R)^2\right]^{3/2}\right] + 3 (c-R)\left[x_n\left(R_o^2 - x_n^2\right)^{1/2}\right.\right.$$

$$\left.\left. + (c-R)\left[R_o^2 - (c-R)^2\right]^{1/2} + R_o^2 \ (\arcsin(x_n/R_o) + \arcsin[(c-R)/R_o])\right]\right\} \tag{4.29}$$

$$C_s = V_1 + V_2 \tag{4.30}$$

$$T_s = V_{3a} + V_{3b} \tag{4.31}$$

$$V_1x_1 = (k_1/12) \left\{ 3R_o x_n \left(R_o^2 - x_n^2\right)^{1/2} + 3R_o^2(c-R)\left[R_o^2 - (c-R)^2\right]^{1/2} + 3R_o^4\{\arcsin(x_n/R_o)\right.$$

$$+ \arcsin[(c-R)/Ro]\} - 6x_n\left(R_o^2 - x_n^2\right)^{3/2} - 6(c-R)\left[R_o^2 - (c-R)^2\right]^{3/2}$$

$$\left. - 8 (c-R)\left(R_o^2 - x_n^2\right)^{3/2} + 8\left[R_o^2 - (c-R)^2\right]^{3/2}\right\} \tag{4.32}$$

$$C_sx_s = V_1x_1 + V_2x_2 \tag{4.33}$$

$$T_sx_t = V_{3a}x_{3a} + V_{3b}x_{3b} \tag{4.34}$$

Case 4: $c_B < c < (R + R_o)$. $V_1 =$ Equation 4.29, $V_2 =$ Equation 4.11, $V_2x_2 =$ Equation 4.18, $V_1x_1 =$ Equation 4.32, and

$$V_3 = (-k_1/3)\left\{-2\left[R_o^2 - (c-R)^2\right]^{3/2} + 3\,(c-R)\left(-[c-R]\left[R_o^2 - (c-R)^2\right]^{1/2}\right.\right.$$
$$\left.\left. +1.5708R_o^2 - R_o^2\arcsin[(c-R)/R_o]\right)\right\} \tag{4.35}$$

$$C_s = V_1 + V_2 \tag{4.36}$$

$$T_s = V_3 \tag{4.37}$$

$$V_3x_3 = (k_1/24)\left\{3\pi R_o^4 + 12\,(c-R)\left[R_o^2 - (c-R)^2\right]^{3/2} - 6R_o^2\left([c-R]\left[R_o^2 - (c-R)^2\right]^{1/2}\right.\right.$$
$$\left.\left. +R_o^2\arcsin[(c-R)/R_o]\right) - 16(c-R)\left[R_o^2 - (c-R)^2\right]^{3/2}\right\} \tag{4.38}$$

$$C_sx_s = V_1x_1 + V_2x_2 \tag{4.39}$$

$$T_sx_t = V_3x_3 \tag{4.40}$$

Case 5: $(R + R_o) < c < c_1$. $V_2 =$ Equation 4.11, $V_3 = 0$, $V_3x_3 = 0$, $V_2x_2 =$ Equation 4.18, $T_s = 0$, $T_sx_t = 0$, and

$$c_1 = Re_c/(e_c - e_s) \tag{4.41}$$

$$V_1 = (k_1/3)\left\{-2\left(R_o^2 - x_n^2\right)^{3/2} + 3(c-R)\left[x_n\left(R_o^2 - x_n^2\right)^{1/2}\right.\right.$$
$$\left.\left. +R_o^2\arcsin(x_n/R_o) + 1.5708R_o^2\right]\right\} \tag{4.42}$$

$$C_s = V_1 + V_2 \tag{4.43}$$

$$V_1x_1 = (k_1/24)\left\{-12x_n\left(R_o^2 - x_n^2\right)^{3/2} + 6R_o^2\left[x_n\left(R_o^2 - x_n\right)^{1/2} + R_o^2\arcsin(x_n/R_o)\right]\right.$$
$$\left. + 3\pi R_o^4 - 16(c-R)\left(R_o^2 - x_n^2\right)^{3/2}\right\} \tag{4.44}$$

$$C_sx_s = V_1x_1 + V_2x_2 \tag{4.45}$$

Case 6: $c_1 < c < c_2$. $V_2 =$ Equation 4.11, $V_2x_2 =$ Equation 4.18, $V_1 =$ Equation 4.42, $V_1x_1 =$ Equation 4.44, $V_3 = 0$, $V_3x_3 = 0$, $T_s = 0$, $T_sx_t = 0$, and

$$c_2 = [e_c/(e_c - e_s)](R + R_o) \tag{4.46}$$

$$C_s = V_1 + V_2 \tag{4.47}$$

$$C_s x_s = V_1 x_1 + V_2 x_2 \tag{4.48}$$

Case 7: When $c > c_2$. $V_1 = 0$, $V_1 x_1 = 0$, $V_3 = 0$, $V_3 x_3 = 0$, $T_s = 0$, $C_s x_s = 0$, $T_s x_t = 0$, and

$$V_2 = k_2 \, \pi R_o^2 \tag{4.49}$$

$$C_s = V_2 \tag{4.50}$$

To write the set of equations for the inner circular section, replace R_o by R in Equations 4.5 to 4.50. This process will yield another 35 equations. The following are the derived equations for the inner circular section:

Case 1: $0 < c < c_o$.

$$V_2 = k_2 \left\{ 1.5708 R^2 - \left[x_n \left(R^2 - x_n^2 \right)^{1/2} + R^2 \, \arcsin\, (x_n/R) \right] \right\} \tag{4.51}$$

$$V_1 = (k_1/3) \left\{ 3 \, [2Rc - c^2]^{3/2} - \left(R^2 - x_n^2 \right)^{3/2} + 3(c-R) \left[x_n \left(R^2 - x_n^2 \right)^{1/2} \right. \right.$$
$$\left. \left. + R^2 \, \arcsin(x_n/R) \right] - (R-c)[2Rc - c^2]^{1/2} + R^2 \, \arcsin[(R-c)/R] \right\} \tag{4.52}$$

$$V_{3a} = -(k_1/3) \left\{ -2[2Rc - c^2]^{3/2} - 3(c-R) \left[x_p \left[R^2 - x_p^2 \right]^{1/2} + R^2 \, \arcsin(x_p/R) \right] \right\} \tag{4.53}$$

$$V_{3b} = k_2 \left\{ x_p \left(R^2 - x_p^2 \right)^{1/2} + R^2[1.5708 + \arcsin(x_p/R)] \right\} \tag{4.54}$$

$$C_s = V_1 + V_2 \tag{4.55}$$

$$T_s = V_{3a} + V_{3b} \tag{4.56}$$

$$V_1 x_1 = (k_1/12) \left\{ -6 \, x_n \left[R^2 - x_n^2 \right]^{3/2} + 3R^2 \left[x_n \left(R^2 - x_n^2 \right)^{1/2} + R^2 \, \arcsin\, (x_n/R) \right] \right.$$
$$+ 6(R-c)([2Rc - c^2]^{3/2} - 3R^2[(R-c)[2Rc - c^2]^{1/2} + R^2 \, \arcsin\, [(R-c)/R]])$$
$$\left. - 8 \, (c-R) \left(\left[R^2 - x_n^2 \right]^{3/2} - [2Rc - c^2]^{3/2} \right) \right\} \tag{4.57}$$

$$V_2 x_2 = (2k_2/3)\left(R^2 - x_n^2\right)^{3/2} \tag{4.58}$$

$$
\begin{aligned}
V_{3a} x_{3a} = (k_1/12) \Big\{ &\left(R^2 - x_p^2\right)^{3/2} (6x_p + 8[c - R]) - [2Rc - c^2]^{3/2}\,(2c - 2R) \\
&+ 3R^2(-x_p \left[R^2 - x_p^2\right]^{1/2} + R^2 \ \arcsin(x_p/R) + (R - c)[2Rc - c^2]^{1/2} \\
&+ R^2 \arcsin[(R - c)/R] \Big)\Big\}
\end{aligned}
\tag{4.59}
$$

$$V_{3b} x_{3b} = (2k_2/3)\left(R^2 - x_p^2\right)^{3/2} \tag{4.60}$$

$$C_s x_s = V_1 x_1 + V_2 x_2 \tag{4.61}$$

$$T_s x_t = V_3 x_{3a} + V_{3b} x_{3b} \tag{4.62}$$

Case 2: $c_o < c < R$. V_2 = Equation 4.51, $V_2 x_2$ = Equation 4.58, and $V_{3b} x_{3b}$ = Equation 4.60.

$$
\begin{aligned}
V_1 = (k_1/3) \Big\{ &-2\left(\left[R^2 - x_n^2\right]^{3/2} - \left[R_o^2 - (c - R)^2\right]^{1/2}\right) + 3(c - R)\left(x_n[R^2 - x_n^2]^{1/2}\right. \\
&+ (c - R)[2Rc - c^2]^{1/2} + R^2 \ \arcsin\ (x_n/R) + \arcsin[(c - R)/R]) \Big\}
\end{aligned}
\tag{4.63}
$$

$$
\begin{aligned}
V_1 x_1 = (k_1/12) \Big\{ &3R^2 x_n \left(R^2 - x_n^2\right)^{1/2} + 3R^2\ (c - R)[2Rc - c^2]^{1/2} + 3R^4\ (\arcsin(x_n/R) \\
&+ \arcsin[(c - R)/R]) - 6x_n \left(R^2 - x_n^2\right)^{3/2} - 6(c - R)[2Rc - c^2]^{3/2} \\
&- 8(c - R)\left(R^2 - x_n^2\right)^{3/2} + 8(c - R)[2Rc - c^2]^{3/2} \Big\}
\end{aligned}
\tag{4.64}
$$

$$
\begin{aligned}
V_{3a} = -(k_1/3) \Big\{ &-2[2Rc - c^2]^{3/2} + 3(c - R)([R - c][2Rc - c^2]^{1/2} + R^2 \ \arcsin[(R - c)/R] \\
&+ 2\left[R^2 - x_p^2\right]^{3/2} + 3(c - R)\left(x_p\left[R^2 - x_p^2\right]^{1/2} + R^2 \ \arcsin[x_p/R]\right) \Big\}
\end{aligned}
\tag{4.65}
$$

$$V_{3b} = K_2 \left\{ -x_p \left(R^2 - x_p^2\right)^{1/2} + R^2[1.5708 - \mathrm{arcisn}\,(x_p/R)] \right\} \tag{4.66}$$

$$V_{3a}x_{3a} = (k_1/12)\left\{\left(R^2 - x_p^2\right)^{3/2}(6x_p - 8[c-R]) + [2Rc - c^2)]^{3/2}(2c - 2R)\right.$$

$$- 3R^2\left(x_p\left[R^2 - x_p^2\right]^{1/2} + R^2 \arcsin[x_p/R]\right) + (R-c)([2Rc - c^2]^{1/2}$$

$$\left. + R^2 \arcsin[(R-c)/R]\right\} \tag{4.67}$$

Case 3: $R < c < C_B$. $V_2 =$ Equation 4.51, $V_2x_2 =$ Equation 4.58, $V_{3a} =$ Equation 4.65, $V_{3ax3a} =$ Equation 4.67, $V_{3b} =$ Equation 4.66, $V_{3b}x_{3b} =$ Equation 4.60, and $C_B = [e_c/(e_c + e_s)](R + R_o)$.

$$V_1 = (k_1/3)\left\{-2\left(\left(2R^2 - x_n^2\right)^{3/2} - [2Rc - c^2]^{3/2}\right) + 3(c-R)x_n\left(R^2 - x_n^2\right)^{1/2}\right.$$

$$\left. + (c-R)([2Rc - c^2]^{1/2} + R^2(\arcsin(x_n/R) + \arcsin[(c-R)/R])\right\} \tag{4.68}$$

$$C_s = V_1 + V_2 \tag{4.69}$$

$$T_s = V_{3a} + V_{3b} \tag{4.70}$$

$$V_1x_1 = (k_1/12)\left\{3Rx_n\left(R^2 - x_n^2\right)^{1/2} + 3R^2(c-R)[2Rc - c^2]^{1/2} + 3R^4\{\arcsin(x_n/R)\right.$$

$$+ \arcsin[(c-R)/R]\} - 6x_n\left(R^2 - x_n^2\right)^{3/2} - 6(c-R)[2Rc - c^2]^{3/2}$$

$$\left. - 8(c-R)\left(R^2 - x_n^2\right)^{3/2} 8[2Rc - c^2]^{3/2}\right\} \tag{4.71}$$

$$C_s x_s = V_1 x_1 + V_2 x_2 \tag{4.72}$$

$$T_s x_t = V_{3a}x_{3a} + V_{3b}x_{3b} \tag{4.73}$$

Case 4: $C_B < c < (R + R_o)$. $V_1 =$ Equation 4.68, $V_2 =$ Equation 4.51, $V_2x_2 =$ Equation 4.58, $V_1x_1 =$ Equation 4.71, and

$$V_3 = (-k_1/3)\{-2[2Rc - c^2]^{3/2} + 3 (c-R)(-[c-R][2Rc - c^2]^{1/2}$$

$$+ 1.5708R^2 - R^2\arcsin[(c-R)/R])\} \tag{4.74}$$

$$C_s = V_1 + V_2 \tag{4.75}$$

$$T_s = V_3 \tag{4.76}$$

$$V_3x_3 = (k_1/24)\{3\pi R^4 + 12 (c-R)[2Rc - c^2]^{3/2} - 6R^2 ([c-R][2Rc - c^2]^{1/2}$$

$$+ R^2 \arcsin[(c-R)/R]) - 16 (c-R)[2Rc - c^2]^{3/2}\} \tag{4.77}$$

$$C_s x_s = V_1 x_1 + V_2 x_2 \tag{4.78}$$

$$T_s x_t = V_3 x_3 \tag{4.79}$$

Case 5: $(R + R_o) < c < C_1$. $V_2 =$ Equation 4.51, $V_3 = 0$, $V_3 x_3 = 0$, $V_2 x_2 =$ Equation 4.58, $T_s = 0$, $T_s x_t = 0$, and

$$C_1 = Re_c/(e_c - e_s) \tag{4.80}$$

$$V_1 = (k_1/3)\left\{-2\left(R^2 - x_n^2\right)^{3/2} + 3(c - R)\left[x_n\left(R^2 - x_n^2\right)^{1/2}\right.\right.$$
$$\left.\left. + R^2 \arcsin(x_n/R) + 1.5708 R^2\right]\right\} \tag{4.81}$$

$$C_s = V_1 + V_2 \tag{4.82}$$

$$V_1 x_1 = (k_1/24)\left\{-12x_n\left(R^2 - x_n^2\right)^{3/2} + 6R_0^2\left[x_n\left(R_0^2 - x_n\right)^{1/2} + R^2 \arcsin(x_n/R)\right]\right.$$
$$\left. + 3\pi R^4 - 16(c - R)(R^2 - x_n^2)^{3/2}\right\} \tag{4.83}$$

$$C_s x_s = V_1 x_1 + V_2 x_2 \tag{4.84}$$

Case 6: $C_1 < c < C_2$. $V_2 =$ Equation 4.47, $V_2 x_2 =$ Equation 4.54, $V_1 =$ Equation 4.77, $V_1 x_1 =$ Equation 4.79, $V_3 = 0$, $V_3 x_3 = 0$, $T_s = 0$, $T_s x_t = 0$, and

$$C_2 = [e_c/(e_c - e_s)](R + R_o) \tag{4.85}$$

$$C_s = V_1 + V_2 \tag{4.86}$$

$$C_s x_s = V_1 x_1 + V_2 x_2 \tag{4.87}$$

Case 7: $c > C_2$. $V_1 = 0$, $V_1 x_1 = 0$, $V_3 = 0$, $V_3 x_3 = 0$, $T_s = 0$, $C_s x_s = 0$, $T_s x_t = 0$, and

$$V_2 = k_2 \pi R_0^2 \tag{4.88}$$

$$C_s = V_2 \tag{4.89}$$

Equations 4.1 to 4.89 are for steel forces and can be programmed in an Excel worksheet to obtain the numerical values of steel capacity. The equations for concrete forces were derived in Chapter 3. The concrete forces and the steel forces are added together to obtain the ultimate capacity of a concrete-filled steel pipe.

Since all parameters are represented in the derived equations, all that is required is to substitute the geometry of the pipe and the concrete and steel stress in the Excel spreadsheets to obtain the capacity of any steel pipe filled with concrete.

4.2.2 Steel forces for rectangular tubing

Steel tubular columns can be hollow or concrete filled. For hollow tubular sections, the steel forces are covered in Chapter 2, but we cannot use the equations for hollow steel tubular columns for CFT columns. Figure 4.3 shows a rectangular concrete-filled steel tubular section in which the concrete core was allowed to develop to its ultimate strength. In this condition of stress and strain in the concrete, the steel shell reaches the yield stress since steel strain is less than that of concrete. The steel forces are now a function of the concrete strain where the strain line of the section is pivoted.

To implement the analytical method, all variable parameters should be accounted for in the analysis. These variables are the dimensions of the rectangular steel tubular section and the ultimate and yield stress of concrete and steel, respectively. The rectangular steel tubular section can then be drawn and these parameters labeled.

To start the analysis, a line is drawn through the center of the section, with an inclination anywhere between the horizontal axis and the vertical axis. This line is designated as the capacity axis. For practical purposes, this line should lie along the diagonal of the rectangular section as a limit since most resultant external load will lie within the sector defined by the horizontal axis and the diagonal.

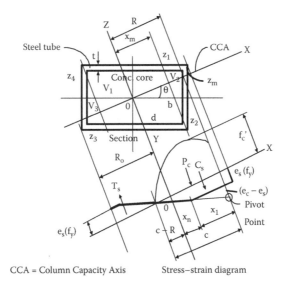

CCA = Column Capacity Axis   Stress–strain diagram

Figure 4.3 Rectangular CFT steel tubular column.
Source: Jarquio, Ramon V. *Analytical Method in Reinforced Concrete.* Universal Publishers, Boca Raton, FL. p. 107. Reprinted with Permission.

The line perpendicular to this capacity axis will be designated as the moment axis. The values of bending moments obtained will be parallel to the capacity axis and represent the component of the resultant moment. To obtain the resultant moment capacity of the section, moments around the capacity axis are also calculated. The square root of the sum of the squares of these moments is the value of the resultant bending moment for the section's biaxial bending capacity.

Hundreds of equations are required to complete the expressions for the forces and moments that the section can develop at its ultimate strength conditions. The author, in his first book, *Analytical Method in Reinforced Concrete*, has shown the equations for CFT columns using his modified stress-strain diagram for steel. Readers may refer to this book for the technique used in analyzing CFT columns as there are no other references available in standard literature using the analytical method for determining the capacities of these columns. Only by using the results of the analytical method can a comparison be made to expose the crudeness of the current one-line equation standard interaction formula for biaxial bending.

The dimensions of a rectangular steel tubular section are the width b and depth d of the rectangular concrete core. The thickness of the shell is designated as t. The column capacity axis is the X-axis, and the moment axis is the Z-axis. The inclination of the capacity axis is designated as θ from the horizontal. Draw lines perpendicular to the X-axis and passing through the corners and the center of the outer rectangle. These will divide the section into three main stress volumes designated as V_1, V_2, and V_3 and their corresponding bending moments V_1x_1, V_2x_2, and V_3x_3. Designate the sides of the rectangle as lines z_1, z_2, z_3, and z_4. Designate the ordinate of the outer rectangle as z_m and the abscissa as $h/2$. Using the point–slope form of the straight line from analytic geometry, write the equations of the sides of the rectangle.

In the XY plane draw the stress and strain for steel. The equation of the steel stress–strain diagram is that of a straight line and can be easily written as Equation 4.6, 4.7 or 4.8. Draw this line for a typical compressive depth and designate its distance from the compressive edge as c.

The intersections of the steel stress diagram with the lines from the corners and center of the rectangular section will define the limits of the different boundaries of the stress volumes and thus the forces and moments to be calculated. There will be several envelopes of values for c to be calculated as a result.

We shall be using the principle of superposition. First obtain the equations for the outer rectangular section and then obtain the equations for the inner rectangular section. The steel forces and moments are obtained by subtracting the inner values from the outer values. For concrete forces and moments refer to Chapter 3.

The concrete-filled rectangular steel tubular column has notations similar to those in Chapter 2, except for the stress volume notations plus the concrete core inside the tube. Here, the stress diagram is that of the CRSI (Concrete Reinforcing Steel Institute), wherein the compressive and tensile steel stresses are mostly in yield conditions when the compressive depth of concrete c is less than that at balanced conditions. This is different from the author's published solution in *Analytical Method in Reinforced Concrete*, in which the tensile stress at the farthest point from the neutral axis is held at the value of the yield stress and decreases as c is decreased from the balanced condition.

First determine the equations for the steel forces and moments applicable to the outer rectangular section. This procedure will yield the steel forces and bending moments that can be developed at the ultimate condition of the concrete core. The limiting concrete strain of 0.003 is the reference for corresponding steel strains in the column section. The steel forces and moments are then computed. The same procedure is followed for the inner rectangular steel section.

This ultimate strength capacity is represented by a curve of plotted values from the Excel spreadsheets. The vertical scale is for the axial capacity, P, and the horizontal scale is for the bending moment capacity, M, of the column section. The eccentricity, $e = M/P$ represents the displacement of the column section with respect to the plumb line because of buckling of the column length. When buckling is not present, as in a stub column, the eccentricity represents the equivalent position of the axial load from the plumb line.

The capacity curve shows the dependence of the axial capacity with the bending moment; that is, when the bending moment is increased, the axial capacity is decreased. When the axial capacity is increased, the bending moment is decreased. The accidental eccentricity due to the column section itself being out of plumb, plus deviation of the application of the external axial and bending moment loads at supports, adds an external bending moment acting on the section. When this occurs, the internal capacity of the section, that is, the axial capacity, should also be reduced. For this reason, the final external axial and bending moment loads determined by the structural engineer are the pair of values to be plotted on the capacity curve to determine the section's ability to support these loads.

The equations for the outer rectangular section are derived first, followed by the inner rectangular section. From analytic geometry, the equations for the sides of the outer rectangular column section are similar to those shown in the previous chapters. These are listed here for completeness of the equations used for solving the rectangular CFT column section:

$$z_1 = - \tan q \, (x - R_o) + z_m \qquad (4.90)$$

$$z_2 = \cot q \, (x - R_o) + z_m \qquad (4.91)$$

$$z_3 = - \tan q \, (x + R_o) - z_m \qquad (4.92)$$

$$z_4 = \cot q \, (x + R_o) - z_m \qquad (4.93)$$

in which

$$R_o = (1/2)\{(d + 2t) \cos \theta + (b + 2t) \sin \theta\} \qquad (4.94)$$

$$z_m = (1/2)\{(b + 2t) \cos \theta - (d + 2t) \sin \theta\} \quad \text{when} \quad \theta < [(\pi/2) - \alpha] \quad (4.95)$$

$$z_m = (1/2)\{(d + 2t) \sin \theta - (b + 2t) \cos \theta\} \quad \text{when} \quad \theta > [(\pi/2) - \alpha] \quad (4.96)$$

in which

$$\alpha = \arctan(b/d) \tag{4.97}$$

The equation of the triangular steel stress is given by

$$y = k\,\{x + (c - R)\} \tag{4.98}$$

in which $k = k_1$ or k_3, depending on the position of c from the compressive concrete edge of the CFT column section and the balanced condition (simultaneous development of ultimate concrete and steel stress).

$$k_1 = 29000e_c/c \tag{4.99}$$

The equation for the uniform steel stress is given by

$$y = k_2 = f_y \tag{4.100}$$

The intersection of uniform stress with triangular stress is given by

$$x_1 = (1/e_c)c(e_c - e_s) \tag{4.101}$$

$$x_n = R - x_1 \tag{4.102}$$

$$R = (1/2)[d \cos \theta - b \sin \theta] \tag{4.103}$$

For limits of pressure volume,

$$x_m = (1/2)\{(d + 2t)\cos \theta - (b + 2t) \sin \theta\} \tag{4.104}$$

in which e_c = concrete strain, f_y = steel yield stress, t = wall thickness, θ = the inclination of the column capacity axis with respect to the horizontal axis, b = the width of the rectangular section, and d = the length or depth of the rectangular section.

Using calculus, V_1, V_2, and V_3 are evaluated within their respective limits defined by the limiting envelopes of c from the compressive concrete edge. These envelopes are influenced by the limits of the strain diagram for uniform and triangular shape. Figures 4.4 and 4.5 show the stress–strain variables to be considered in the analysis.

The integration process should yield 116 equations for the outer rectangular section and another 116 equations for the inner rectangular section. The difference in values between the outer and inner sections is the measure of the steel forces. Add the concrete forces to this to obtain the ultimate strength of the CFT column.

The ultimate strength capacity when $\theta = \alpha$ is recommended to resist biaxial bending since the capacity is less than at other positions of θ. Plot

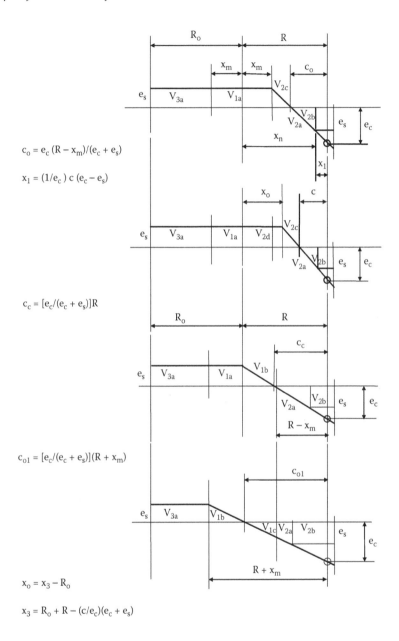

$$c_o = e_c (R - x_m)/(e_c + e_s)$$

$$x_1 = (1/e_c) \, c \, (e_c - e_s)$$

$$c_c = [e_c/(e_c + e_s)]R$$

$$c_{o1} = [e_c/(e_c + e_s)](R + x_m)$$

$$x_o = x_3 - R_o$$

$$x_3 = R_o + R - (c/e_c)(e_c + e_s)$$

Figure 4.4 Stress–strain variables.

the values of P and M to yield the ultimate strength capacity curve of the CFT column section.

Case 1: $0 < c < c_o$.

$$V_1a = 4k_2k_3x_m \qquad (4.105)$$

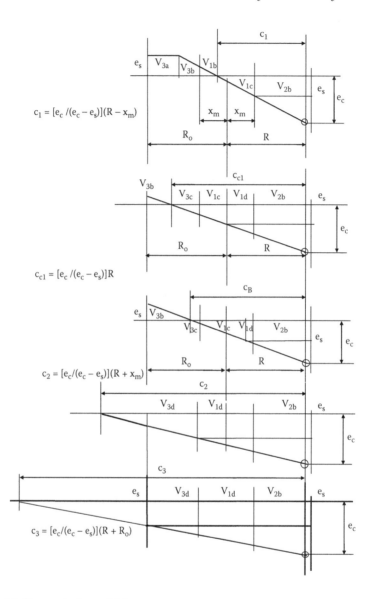

Figure 4.5 Stress–strain variables.

$$V_{2a} = -(k_1/6)(k_4)\left\{x_n\left[2x_n^2 + 3(c-R-R_o)x_n - 6R_o(c-R)\right] - (c-R)^2(c+3R_o-R)\right\}$$

(4.106)

$$V_{2b} = (k_2/2)(k_4)\left\{R_o^2 + x_n(x_n - 2R_o)\right\}$$

(4.107)

$$V_{2c} = (k_3/6)(k_4)\left\{(R-c)^2[c+3R_o-R] - x_o 2x_o^2 + 3x_o(c-R-R_o) - 6R_o(c-R)\right\}$$

(4.108)

$$V_{2d} = -(k_2/2)(k_4)\{x_o(x_o - 2R_o) - x_m(x_m - 2R_o)\} \qquad (4.109)$$

$$V_{3a} = (k_2 k_4/2)\left\{R_o^2 + x_m(x_m - 2R_o)\right\} \qquad (4.110)$$

$$C_s = V_{2a} + V_{2b} \qquad (4.111)$$

$$T_s = V_{1a} + V_{3a} + V_{2c} \qquad (4.112)$$

$$V_{1a}x_{1a} = 0 \qquad (4.113)$$

$$V_{2a}x_{2a} = -(k_1/12)(k_4)\left\{x_n^2\left[3x_n^2 + 4(c - R - R_o)x_n - 6R_o(c - R)\right]\right.$$
$$\left. - (R - c)^3[c + 2R_o - R]\right\} \qquad (4.114)$$

$$V_{2b}x_{2b} = (k_2/6)(k_4)\left\{R_o^3 + \left[2x_n^3 - 3R_o x_n^2\right]\right\} \qquad (4.115)$$

$$V_{2c}x_{2c} = (k_3/12)(k_4)\left\{(R - c)^3[c + 2R_o - R] - x_o^2\left[3x_o^2 + 4x_o(c - R - R_o)6R_o(c - R)\right]\right\} \qquad (4.116)$$

$$V_{2d}x_{2d} = -(k_2/6)(k_4)\left\{x_o^2(2x_o - 3R_o) - x_m^2(2x_m - 3R_o)\right\} \qquad (4.117)$$

$$V_{3a}x_{3a} = (k_3/12)(k_4)\left\{x_m^2[3R_o - 2x_m] - R_o^3\right\} \qquad (4.118)$$

$$C_s x_s = V_{2a}x_{2a} + V_{2b}x_{2b} \qquad (4.119)$$

$$T_s x_t = V_{1a}x_{1a} - V_{2c}x_{2c} + V_{3a}x_{3a} - V_{2d}x_{2d} \qquad (4.120)$$

Case 2: $c_o < c < (R - x_m)$. V_{2b} = Equation 4.107, $V_{2b}x_{2b}$ = Equation 4.115, V_{3a} = Equation 4.110, $V_{3a}x_{3a}$ = Equation 4.118, V_{2a} = Equation 4.106, $V_{2a}x_{2a}$ = Equation 4.114, and

$$c_o = e_c[R - x_m]/(e_c + e_s) \qquad (4.121)$$

$$k_4 = (\cot\theta + \tan\theta) \qquad (4.122)$$

$$k_3 = R_o \tan\theta + z_m \qquad (4.123)$$

$$V_{1a} = 2k_2 k_3(x_o + x_m) \qquad (4.124)$$

$$V_{1b} = -(k_1 k_3)\left\{x_m^2 - x_o^2 + 2(c - R)(x_m - x_o)\right\} \qquad (4.125)$$

$$V_{2c} = (k_1/6)(k_4)\left\{(R-c)^2(3R_o + c - R) - x_m\left[2x_m^2 + 3x_m(c - R - R_o) - 6R_o(c - R)\right]\right\}$$

(4.126)

$$C_s = V_{2a} + V_{2b}$$

(4.127)

$$T_s = V_{3a} + V_{1a} + V_{1b} + V_{2c}$$

(4.128)

$$V_{1a}x_{1a} = -(k_2k_3)\left(x_o^2 - x_m^2\right)$$

(4.129)

$$V_{1b}x_{1b} = (k_1/3)(k_3)\left\{2\left(x_m^3 - x_o^3\right) + 3(c - R)\left(x_m^2 - x_o^2\right)\right\}$$

(4.130)

$$V_{2ac2c} = (k_1/12)(k_4)\left\{(R-c)^3(2R_o + c - R) - x_m^2\left[2x_m^2 + 4x_m(c - R - R_o)\right.\right.$$
$$\left.\left. -6R_o(c - R)\right]\right\}$$

(4.131)

$$C_sx_s = V_{2a}x_{2a} + V_{2b}x_{2b}$$

(4.132)

$$T_sx_t = V_{3a}x_{3a} + V_{1a}x_{1a} - V_{1b}x_{1b} - V_{2c}x_{2c}$$

(4.133)

Case 3: $(R - x_m) < c < c_c$. V_{2b} = Equation 4.107, $V_{2b}x_{2b}$ = Equation 4.115, V_{3a} = Equation 4.110, $V_{3a}x_{3a}$ = Equation 4.118, V_{2a} = Equation 4.106, $V_{2a}x_{2a}$ = Equation 4.114, V_{1a} = Equation 4.124, $V_{1a}x_{1a}$ = Equation 4.129, and

$$c_c = [e_c/(e_c + e_s)]R$$

(4.134)

$$V_{1b} = (k_1k_3)\{(R - c)^2 + x_o[x_o + 2(c - R)]\}$$

(4.135)

$$V_{1c} = (k_1k_3)\{(R - c)^2 + x_m[x_m + 2(c - R)]\}$$

(4.136)

$$C_s = V_{2a} + V_{2b} + V_{1c}$$

(4.137)

$$T_s = V_{3a} + V_{1a} + V_{1b}$$

(4.138)

$$V_{1b}x_{1b} = (k_1/3)(k_3)\left\{(R-c)^3 + x_o^2\left[2x_o + x_o^2[2x_o + 3](c - R)\right]\right\}$$

(4.139)

$$V_{1c}x_{1c} = (k_1/3)(k_3)\left\{x_m^2[2x_m + 3(c - R)] + (R - c)^3\right\}$$

(4.140)

$$C_sx_s = V_{2a}x_{2a} + V_{2b}x_{2b} + V_{1c}x_{1c}$$

(4.141)

$$T_sx_t = V_{3a}x_{3a} + V_{1a}x_{1a} + V_{1b}x_{1b}$$

(4.142)

Case 4: $c_c < c < c_{o1}$. V_{2b} = Equation 4.107, $V_{2b}x_{2b}$ = Equation 4.115, V_{3a} = Equation 4.110, $V_{3a}x_{3a}$ = Equation 4.118, V_{2a} = Equation 4.106, $V_{2a}x_{2a}$ = Equation 4.114, V_{1c} = Equation 4.136, $V_{1c}x_{1c}$ = Equation 4.140, $V_{1a}x_{1a}$ = Equation 4.129, and

$$c_{o1} = [e_c/(e_c + e_s)](R + x_m) \tag{4.143}$$

$$V_{1a} = -2k_2k_3(x_o - x_m) \tag{4.144}$$

$$V_{1b} = (k_1k_3)\{(R - c)^2 + x_o[x_o - 2(c - R)]\} \tag{4.145}$$

$$C_s = V_{2a} + V_{2b} + V_{1c} \tag{4.146}$$

$$T_s = V_{3a} + V_{1a} + V_{1b} \tag{4.147}$$

$$V_{1b}x_{1b} = -(k_1/3)(k_3)\{(R - c)^3 + x_o^2[-2x_o + 3(c - R)]\} \tag{4.148}$$

$$C_sx_s = V_{2a}x_{2a} + V_{2b}x_{2b} + V_{1c}x_{1c} \tag{4.149}$$

$$T_sx_t = V_{3a}x_{3a} + V_{1a}x_{1a} + V_{1b}x_{1b} \tag{4.150}$$

Case 5: $c_{o1} < c < c_1$. V_{2b} = Equation 4.107, $V_{2b}x_{2b}$ = Equation 4.115, V_{3a} = Equation 4.110, V_{1c} = Equation 4.136, $V_{1c}x_{1c}$ = Equation 4.140, and

$$c_1 = [e_c/(e_c - e_s)](R - x_m) \tag{4.151}$$

$$V_{3a} = (k_2/4)(k_4)\{r_o^2 + x_o(x_o + 2R_o)\} \tag{4.152}$$

$$V_{3b} = -(k_1/6)(k_4)\{-2(x_m^3 + x_o^3) + 3(R_o + c - R)(x_m^2 - x_o^2) - 6R_o^2(c - R)(x_m + x_o)\} \tag{4.153}$$

$$V_{1b} = (k_1k_3)\{(R - c)^2 + x_m[x_m - 2(c - R)]\} \tag{4.154}$$

$$C_s = V_{2a} + V_{2b} + V_{1c} \tag{4.155}$$

$$T_s = V_{3a} + V_{1a} + V_{1b} \tag{4.156}$$

$$V_{3a}x_{3a} = -(k_2/6)(k_4)\{x_o^2(2x_o + 3R_o) - R_o^3\} \tag{4.157}$$

$$V_{3b}x_{3b} = (k_1/12)(k_4)\{3(x_m^4 - x_o^4) - 4(R_o + c - R)(x_m^3 + x_o^3) + 6R_o(c - R)(x_m^2 - x_o^2)\} \tag{4.158}$$

$$V_{1b}x_{1b} = -(k_1/3)(k_3)\{(R-c)^3 + x_m^2[-2x_m + 3(c-R)]\}$$ (4.159)

$$C_sx_s = V_{2a}x_{2a} + V_{2b}x_{2b} + V_{1c}x_{1c}$$ (4.160)

$$T_sx_t = V_{3a}x_{3a} + V_{1a}x_{1a} + V_{1b}x_{1b}$$ (4.161)

Case 6: $c_1 < c < (R + x_m)$. V_{3a} = Equation 4.152, $V_{3a}x_{3a}$ = Equation 4.157, V_{3b} = Equation 4.153, $V_{3b}x_{3b}$ = Equation 4.158, V_{1b} = Equation 4.154, and $V_{1b}x_{1b}$ = Equation 4.159.

$$V_{1c} = (k_1k_3)\{(R-c)^2 + 2x_n[x_n + 2(c-R)]\}$$ (4.162)

$$V_{1d} = 2(k_2k_3)(x_m - x_n)$$ (4.163)

$$V_{2b} = (k_2/2)(k_4)\{R_0^2 + x_m(x_m - 2R_0)\}$$ (4.164)

$$C_s = V_{2b} + V_{1c} + V_{1d}$$ (4.165)

$$T_s = V_{3a} + V_{3b} + V_{1b}$$ (4.166)

$$V_{1c}x_{1c} = (k_1/3)(k_3)\{x_n^2[2x_n + 3(c-R)] - (c-R)^3\}$$ (4.167)

$$V_{1d}x_{1d} = (k_2k_3)(x_m^2 - x_n^2)$$ (4.168)

$$V_{2b}x_{2b} = (k_2/6)(k_4)\{R_0^3 + x_m^2(2x_m - 3R_0)\}$$ (4.169)

$$C_sx_s = V_{2a}x_{2a} + V_{2b}x_{2b} + V_{1c}x_{1c}$$ (4.170)

$$T_sx_t = V_{3a}x_{3a} + V_{1a}x_{1a} + V_{1b}x_{1b}$$ (4.171)

Case 7: $(R + x_m) < c < c_B$. V_{3a} = Equation 4.152, $V_{3a}x_{3a}$ = Equation 4.157, V_{2b} = Equation 4.164, $V_{2b}x_{2b}$ = Equation 4.169, V_{1d} = Equation 4.163, $V_{1d}x_{1d}$ = Equation 4.168, and

$$C_B = [e_c/(e_c + e_s)](R + R_0)$$ (4.172)

$$V_{3b} = -(k_1/6)(k_4)\{(c-R)^2(c-R-3R_0) - x_o[-2x_o^2 + 3x_o(c-R+R_0) - 6R_0(c-R)]\}$$ (4.173)

$$V_{3c} = (k_1/6)(k_4)\left\{x_m\left[-2x_m^2 + 3x_m(c - R + R_o) - 6R_o(c - R)\right] - (c - R)^2(c - R - 3R_o)\right\}$$

(4.174)

$$V_{1c} = (k_1 k_3)\left\{x_n^2 - x_m^2) + 2(c - R)(x_n + x_m)\right\}$$ (4.175)

$$C_s = V_{2b} + V_{1c} + V_{1d}$$ (4.176)

$$T_s = V_{3a} + V_{3b} + V_{3c}$$ (4.177)

$$V_{3b}x_{3b} = (k_1/12)(k_4)\left\{(c - R)^3(R + 2R_o - c) - x_o^2\left[3x_o^2 - 4x_o(c - R + R_o) + 6R_o(c - R)\right]\right\}$$

(4.178)

$$V_{3c}x_{3c} = -(k_1/12)(k_4)\left\{x_m^2\left[3x_m^2 - 4x_m(c - R + R_o) + 6R_o(c - R)\right] - (c - R)^3(R + 2R_o - c)\right\}$$

(4.179)

$$V_{1c}x_{1c} = -(k_1)(k_3)\left\{2\left(x_n^3 + x_m^3\right) + 3(c - R)\left(x_n^2 - x_m^2\right)\right\}$$ (4.180)

$$C_s x_s = V_{2b}x_{2b} + V_{1c}x_{1c} + V_{1d}x_{1d} - V_{3c}x_{3c}$$ (4.181)

$$T_s x_t = V_{3a}x_{3a} + V_{3b}x_{3b}$$ (4.182)

Case 8: $c_B < c < c_{c1}$. V_{2b} = Equation 4.164, $V_{2b}x_{2b}$ = Equation 4.169, V_{1d} = Equation 4.163, $V_{1d}x_{1d}$ = Equation 4.168, V_{1c} = Equation 4.175, and

$$c_{c1} = [e_c/(e_c - e_s)]R$$ (4.183)

$$V_{3b} = -(k_1/6)(k_4)\left\{(c - R)^2(c - R - 3R_o) - R_o^2(R_o + 3R - 3c)\right\}$$ (4.184)

$$V_{3c} = (k_1/6)(k_4)\left\{x_m\left[-2x_m^2 + 3x_m(c - R + R_o) + 6R_o(c - R)\right](c - R)^2(c - R - 3R_o)\right\}$$

(4.185)

$$C_s = V_{2b} + V_{1c} + V_{1d} + V_{3c}$$ (4.186)

$$T_s = V_{3b}$$ (4.187)

$$V_{3b}x_{3b} = (k_1/12)(k_4)\left\{(c - R)^3(2R_0 + R - c) - R_0^3(2c - 2R - R_0)\right\}$$ (4.188)

$$V_{3c}x_{3c} = -(k_1/12)(k_4)\left\{x_m^2\left[3x_m^2 - 4x_m(R_o + c - R)\right.\right.$$
$$\left.\left. +6R_o(c-R)\right] - (c-R)^3(2R_o + R - c)\right\}$$

(4.189)

$$V_{1c}x_{1c} = (k_1/3)(k_3)\left\{2\left(x_n^3 - x_m^3\right) + 3(c-R)\left(x_n^2 + x_m^2\right)\right\}$$ (4.190)

$$C_s x_s = V_{2b}x_{2b} - V_{1c}x_{1c} + V_{1d}x_{1d} - V_{3c}x_{3c}$$ (4.191)

$$T_s x_t = V_{3b}x_{3b}$$ (4.192)

Case 9: $c_{c1} < c < (R + R_o)$. V_{2b} = Equation 4.164, $V_{2b}x_{2b}$ = Equation 4.169, V_{1d} = Equation 4.163, $V_{1d}x_{1d}$ = Equation 4.168, V_{3b} = Equation 4.184, $V_{3b}x_{3b}$ = Equation 4.188, V_{3c} = Equation 4.185, and $V_{3c}x_{3c}$ = Equation 4.189.

$$V_{1c} = (k_1 k_3)\left\{x_n^2 - x_m^2 - 2(c-R)(x_n - x_m)\right\}$$ (4.193)

$$C_s = V_{2b} + V_{1c} + V_{1d} + V_{3c}$$ (4.194)

$$T_s = V_{3b}$$ (4.195)

$$V_{1c}x_{1c} = -(k_1/3)(k_3)\left\{2\left(x_m^3 - x_n^3\right) + 3(c-R)\left(x_n^2 - x_m^2\right)\right\}$$ (4.196)

$$C_s x_s = V_{2b}x_{2b} - V_{1c}x_{1c} + V_{1d}x_{1d} - V_{3c}x_{3c}$$ (4.197)

$$T_s x_t = V_{3b}x_{3b}$$ (4.198)

Case 10: $(R + R_o) < c < c_2$. V_{2b} = Equation 4.164, $V_{2b}x_{2b}$ = Equation 4.169, $V_{1d}x_{1d}$ = Equation 4.168, and

$$c_2 = [e_c/(e_c - e_s)](R + x_m)$$ (4.199)

$$V_{1d} = 2k_2 k_3(x_m + x_n)$$ (4.200)

$$V_{1c} = (k_1 k_3)\left\{x_n^2 - x_m^2 - 2(c-R)(x_n - x_m)\right\}$$ (4.201)

$$V_{3d} = (k_1/6)(k_4)\left\{x_m\left[-2x_m^2 + 3x_m(c - R + R_o) - 6R_o(c-R)\right] - R_o^2(R_o + 3R - 3c)\right\}$$ (4.202)

$$C_s = V_{2b} + V_{1c} + V_{1d} + V_{3d}$$ (4.203)

$$T_s = 0$$ (4.204)

$$V_{1c}x_{1c} = -(k_1/3)(k_3)\left\{2\left(x_m^3 - x_n^3\right) + 3(c - R)\left(x_n^2 - x_m^2\right)\right\}$$ (4.205)

$$V_{3d}x_{3d} = -(k_1/12)(k_4)\left\{x_m^2\left[3x_m^2 - 4x_m(c - R + R_o)6R_o(c - R)\right] - R_o^3(2c - 2R - R_o)\right\}$$ (4.206)

$$C_s x_s = V_{2b}x_{2b} - V_{1c}x_{1c} + V_{1d}x_{1d} - V_{3d}x_{3d}$$ (4.207)

$$T_s x_t = 0$$ (4.208)

Case 11: $c_2 < c < c_3$. V_{2b} = Equation 4.164, $V_{2b}x_{2b}$ = Equation 4.169, and

$$c_3 = [e_c/(e_c - e_s)](R + R_o)$$ (4.209)

$$V_{1d} = 4k_1k_3x_m$$ (4.210)

$$V_{3d} = (k_2/2)(k_4)\left\{x_m^2 - x_n^2 - 2R_o(x_m - x_n)\right\}$$ (4.211)

$$V_{3b} = (k_1/6)(k_4)\left\{x_n\left[-2x_n^2 + 3x_n(c - R + R_o) - 6R_o(c - R)\right] - R_o^2(R_o + 3R - 3c)\right\}$$ (4.212)

$$C_s = V_{2b} + V_{1d} + V_{3d} + V_{3b}$$ (4.213)

$$V_{1d}x_{1d} = 0$$ (4.214)

$$V_{3d}x_{3d} = -(k_2/6)(k_4)\left\{2\left(x_n^3 - x_m^2\right) + 3R_o\left(x_m^2 - x_n^2\right)\right\}$$ (4.215)

$$V_{3b}x_{3b} = -(k_1/12)(k_4)\left\{x_n^2\left[3x_n^2 - 4x_n(c - R + R_o) + 6R_o(c - R)\right] - R_o^3(2c - 2R - R_o)\right\}$$ (4.216)

$$C_s x_s = V_{2b}x_{2b} - V_{3b}x_{3b} + V_{1d}x_{1d} - V_{3d}x_{3d}$$ (4.217)

Case 12: $c > c_3$. V_{2b} = Equation 4.164, $V_{2b}x_{2b}$ = Equation 4.169, V_{1d} = Equation 4.210, and $V_{1d}x_{1d}$ = Equation 4.214.

$$V_{3d} = (k_2/2)(k_4)\left\{R_o^2 + x_m(x_m - 2R_o)\right\}$$ (4.218)

$$C_s = V_{2b} + V_{1d} + V_{3d}$$ (4.219)

$$V_{3d}x_{3d} = -(k_2/6)(k_4)\left\{x_m^2(-2x_m + 3R_o) - R_o^3\right\}$$ (4.220)

$$C_s x_s = V_{2b}x_{2b} + V_{1d}x_{1d} - V_{3d}x_{3d}$$ (4.221)

Equations 4.105 to 4.221, a total of 116 equations, determine the capacity of the outer rectangular section.

To derive the set of equations for the inner rectangular section of the CFT column, replace x_m with x_2, replace R_o with R, and replace z_m with z_o, in which

$$x_2 = (1/2)[d \cos \theta - b \sin \theta] \qquad (4.222)$$

$$z_o = (1/2)[b \cos \theta - d \sin \theta] \qquad (4.223)$$

These substitutions will generate another 116 equations for the inner rectangular section.

Case 1: $0 < c < c_o$.

$$V_1 a = 4k_2 k_3 x_2 \qquad (4.224)$$

$$V_{2a} = -(k_1/6)(k_4)\left\{x_n\left[2x_n^2 + 3(c - 2R)x_n - 6R(c - R)\right] - (c - R)^2(c + 2R)\right\} \qquad (4.225)$$

$$V_{2b} = (k_2/2)(k_4)\{R^2 + x_n(x_n - 2R)\} \qquad (4.226)$$

$$V_{2c} = (k_3/6)(k_4)\left\{(R - c)^2[c + 2R] - x_o[2x_o^2 + 3x_o(c - 2R) - 6R(c - R)\right\} \qquad (4.227)$$

$$V_{2d} = -(k_2/2)(k_4)\{x_o(x_o - 2R) - x_2(x_2 - 2R)\} \qquad (4.228)$$

$$V_{3a} = (k_2 k_4/2)\{R^2 + x_2 (x_2 - 2R)\} \qquad (4.229)$$

$$C_s = V_{2a} + V_{2b} \qquad (4.230)$$

$$T_s = V_{1a} + V_{3a} + V_{2c} \qquad (4.231)$$

$$V_{1a} x_{1a} = 0 \qquad (4.232)$$

$$V_{2a} x_{2a} = -(k_1/12)(k_4)\left\{x_n^2\left[3x_n^2 + 4(c - 2R)x_n + 6R(c - R)\right] - (R - c)^3[c + R]\right\} \qquad (4.233)$$

$$V_{2b} x_{2b} = (k_2/6)(k_4)\left\{R^3 + \left[2x_n^3 - 3Rx_n^2\right]\right\} \qquad (4.234)$$

$$V_{2c} x_{2c} = (k_3/12)(k_4)\left\{(R - c)^3[c + R] - x_o^2\left[3x_o^2 + 4x_o(c - 2R) - 6R(c - R)\right]\right\} \qquad (4.235)$$

$$V_{2d} x_{2d} = -(k_2/6)(k_4)\left\{x_o^2(2x_o - 3R) - x_m^2(2x_m - 3R)\right\} \qquad (4.236)$$

$$V_{3a}x_{3a} = (k_3/12(k_4))\{x_2^2[3R - 2x_m] - R^3\} \tag{4.237}$$

$$C_s x_s = V_{2a}x_{2a} + V_{2b}x_{2b} \tag{4.238}$$

$$T_s x_t = V_{1a}x_{1a} - V_{2c}x_{2c} + V_{3a}x_{3a} - V_{2d}x_{2d} \tag{4.239}$$

Case 2: $c_o < c < (R - x_2)$. V_{2b} = Equation 4.226, $V_{2b}x_{2b}$ = Equation 4.234, V_{3a} = Equation 4.229, $V_{3a}x_{3a}$ = Equation 4.237, V_{2a} = Equation 4.225, $V_{2a}x_{2a}$ = Equation 4.233, and

$$c_o = e_c[R - x_m]/(e_c + e_s) \tag{4.240}$$

$$k_4 = (\cot\theta + \tan\theta) \tag{4.241}$$

$$k_3 = R\tan\theta + z_o \tag{4.242}$$

$$V_{1a} = 2k_2k_3(x_o + x_2) \tag{4.243}$$

$$V_{1b} = -(k_1k_3)\{x_2^2 - x_o^2 + 2(c - R)(x_2 - x_o)\} \tag{4.244}$$

$$V_{2c} = (k_1/6)(k_4)\{(R - c)^2(2R + c) - x_2[2x_2^2 + 3x_2(c - 2R) - 6R(c - R)]\} \tag{4.245}$$

$$C_s = V_{2a} + V_{2b} \tag{4.246}$$

$$T_s = V_{3a} + V_{1a} + V_{1b} + V_{2c} \tag{4.247}$$

$$V_{1a}x_{1a} = -(k_2k_3)(x_o^2 - x_2^2) \tag{4.248}$$

$$V_{1b}x_{1b} = (k_1/3)(k_3)\{2(x_2^3 - x_o^3) + 3(c - R)(x_2^2 - x_o^2)\} \tag{4.249}$$

$$V_{2ac2c} = (k_1/12)(k_4)\{(R - c)^3(R + c) - x_2^2[2x_2^2 + 4x_2(c - 2R) - 6R(c - R)]\} \tag{4.250}$$

$$C_s x_s = V_{2a}x_{2a} + V_{2b}x_{2b} \tag{4.251}$$

$$T_s x_t = V_{3a}x_{3a} + V_{1a}x_{1a} - V_{1b}x_{1b} - V_{2c}x_{2c} \tag{4.252}$$

Case 3: $(R - x_2) < c < c_c$. V_{2b} = Equation 4.226, $V_{2b}x_{2b}$ = Equation 4.234, V_{3a} = Equation 4.229, $V_{3a}x_{3a}$ = Equation 4.237, V_{2a} = Equation 4.225, $V_{2a}x_{2a}$ = Equation 4.233, V_{1a} = Equation 4.243, $V_{1a}x_{1a}$ = Equation 4.248, and c_c = Equation 4.134.

$$V_{1b} = (k_1k_3)\{(R - c)^2 + x_o[x_o + 2(c - R)]\} \tag{4.253}$$

$$V_{1c} = (k_1 k_3)\{(R - c)^2 + x_2[x_2 + 2(c - R)]\} \tag{4.254}$$

$$C_s = V_{2a} + V_{2b} + V_{1c} \tag{4.255}$$

$$T_s = V_{3a} + V_{1a} + V_{1b} \tag{4.256}$$

$$V_{1b}x_{1b} = (k_1/3)(k_3)\left\{(R - c)^3 + x_o^2[2x_o + x_o^2[2x_o + 3(c - R)]]\right\} \tag{4.257}$$

$$V_{1c}x_{1c} = (k_1/3)(k_3)\left\{x_2^2[2x_2 + 3(c - R)] + (R - c)^3\right\} \tag{4.258}$$

$$C_s x_s = V_{2a}x_{2a} + V_{2b}x_{2b} + V_{1c}x_{1c} \tag{4.259}$$

$$T_s x_t = V_{3a}x_{3a} + V_{1a}x_{1a} + V_{1b}x_{1b} \tag{4.260}$$

Case 4: $c_c < c < c_{o1}$. V_{2b} = Equation 4.226, $V_{2b}x_{2b}$ = Equation 4.234, V_{3a} = Equation 4.229, $V_{3a}x_{3a}$ = Equation 4.237, V_{2a} = Equation 4.225, $V_{2a}x_{2a}$ = Equation 4.232, V_{1c} = Equation 4.254, $V_{1c}x_{1c}$ = Equation 4.258, $V_{1a}x_{1a}$ = Equation 4.248, and

$$c_{o1} = [e_c/(e_c + e_s)](R + x_2) \tag{4.261}$$

$$V_{1a} = -2k_2 k_3 (x_o - x_2) \tag{4.262}$$

$$V_{1b} = (k_1 k_3)\{(R - c)^2 + x_o[x_o - 2(c - R)]\} \tag{4.263}$$

$$C_s = V_{2a} + V_{2b} + V_{1c} \tag{4.264}$$

$$T_s = V_{3a} + V_{1a} + V_{1b} \tag{4.265}$$

$$V_{1b}x_{1b} = -(k_1/3)(k_3)\left\{(R - c)^3 + x_o^2[-2x_o + 3(c - R)]\right\} \tag{4.266}$$

$$C_s x_s = V_{2a}x_{2a} + V_{2b}x_{2b} + V_{1c}x_{1c} \tag{4.267}$$

$$V_{3a}x_{3a} + V_{1a}x_{1a} + V_{1b}x_{1b} \tag{4.268}$$

Case 5: $c_{o1} < c < c_1$. V_{2b} = Equation 4.226, $V_{2b}x_{2b}$ = Equation 4.234, V_{3a} = Equation 4.229, V_{1c} = Equation 4.254, $V_{1c}x_{1c}$ = Equation 4.258, and

$$c_1 = [e_c/(e_c - e_s)](R - x_2) \tag{4.269}$$

$$V_{3a} = (k_2/4)(k_4)\left\{r_o^2 + x_o(x_o + 2R)\right\} \tag{4.270}$$

$$V_{3b} = -(k_1/6)(k_4)\left\{-2\left(x_2^3 + x_o^3\right) + 3c\left(x_2^2 - x_o^2\right) - 6R^2(c - R)(x_2 + x_o)\right\} \quad (4.271)$$

$$V_{1b} = (k_1 k_3)\{(R - c)^2 + x_2[x_2 - 2(c - R)]\} \quad (4.272)$$

$$C_s = V_{2a} + V_{2b} + V_{1c} \quad (4.273)$$

$$T_s = V_{3a} + V_{1a} + V_{1b} \quad (4.274)$$

$$V_{3a}x_{3a} = -(k_2/6)(k_4)\left\{x_0^2(2x_0 + 3R) - R^3\right\} \quad (4.275)$$

$$V_{3b}x_{3b} = (k_1/12)(k_4)\left\{3\left(x_2 - x_0^4\right) - 4(R + c - R)\left(x_2 + x_0^3\right) + 6R(c - R)\left(x_2 - x_0^2\right)\right\} \quad (4.276)$$

$$V_{1b}x_{1b} = -(k_1/3)(k_3)\left\{(R - c)^3 + x_2^2[-2x_2 + 3(c - R)]\right\} \quad (4.277)$$

$$C_s x_s = V_{2a}x_{2a} + V_{2b}x_{2b} + V_{1c}x_{1c} \quad (4.278)$$

$$T_s x_t = V_{3a}x_{3a} + V_{1a}x_{1a} + V_{1b}x_{1b} \quad (4.279)$$

Case 6: $c_1 < c < (R + x_2)$. V_{3a} = Equation 4.270, $V_{3a}x_{3a}$ = Equation 4.275, V_{3b} = Equation 4.271, $V_{3b}x_{3b}$ = Equation 4.276, V_{1b} = Equation 4.272, and $V_{1b}x_{1b}$ = Equation 4.277.

$$V_{1c} = (k_1 k_3)\{(R - c)^2 + 2x_n[x_n + 2(c - R)]\} \quad (4.280)$$

$$V_{1d} = 2(k_2 k_3)(x_2 - x_n) \quad (4.281)$$

$$V_{2b} = (k_2/2)(k_4)\left\{R_0^2 + x_2(x_2 - 2R)\right\} \quad (4.282)$$

$$C_s = V_{2b} + V_{1c} + V_{1d} \quad (4.283)$$

$$T_s = V_{3a} + V_{3b} + V_{1b} \quad (4.284)$$

$$V_{1c}x_{1c} = (k_1/3)(k_3)\left\{x_n^2[2x_n + 3(c - R)] - (c - R)^3\right\} \quad (4.285)$$

$$V_{1d}x_{1d} = (k_2 k_3)\left(x_2^2 - x_n^2\right) \quad (4.286)$$

$$V_{2b}x_{2b} = (k_2/6)(k_4)\left\{R^3 + x_2^2(2x_2 - 3R)\right\} \quad (4.287)$$

$$C_s x_s = V_{2a} x_{2a} + V_{2b} x_{2b} + V_{1c} x_{1c} \qquad (4.288)$$

$$T_s x_t = V_{3a} x_{3a} + V_{1a} x_{1a} + V_{1b} x_{1b} \qquad (4.289)$$

Case 7: $(R + x_m) < c < c_B$. V_{3a} = Equation 4.270, $V_{3a} x_{3a}$ = Equation 4.275, V_{2b} = Equation 4.282, $V_{2b} x_{2b}$ = Equation 4.287, V_{1d} = Equation 4.281, $V_{1d} x_{1d}$ = Equation 4.286, and c_B = Equation 4.172.

$$V_{3b} = -(k_1/6)(k_4)\left\{ (c - R)^2 (c - 4R) - x_0 \left[-2x_0^2 + 3x_0 c - 6R_0 (c - R) \right] \right\} \qquad (4.290)$$

$$V_{3c} = (k_1/6)(k_4)\left\{ x_2 \left[-2x_2^2 + 3x_2 c - 6R(c - R) \right] - (c - R)^2 (c - 4R) \right\} \qquad (4.291)$$

$$V_{1c} = (k_1 k_3)\left\{ x_n^2 - x_m^2 + 2(c - R)(x_n + x_m) \right\} \qquad (4.292)$$

$$C_s = V_{2b} + V_{1c} + V_{1d} \qquad (4.293)$$

$$T_s = V_{3a} + V_{3b} + V_{3c} \qquad (4.294)$$

$$V_{3b} x_{3b} = (k_1/12)(k_4)\left\{ (c - R)^3 (3R - c) - x_0^2 \left[3x_0^2 - 4x_0 c + 6R(c - R) \right] \right\} \qquad (4.295)$$

$$V_{3c} x_{3c} = -(k_1/12)(k_4)\left\{ x_2^2 \left[3x_2^2 - 4x_2 + 6R(c - R) \right] - (c - R)^3 (3R - c) \right\} \qquad (4.296)$$

$$V_{1c} x_{1c} = -(k_l)(k_3)\left\{ 2\left(x_n^3 + x_2^3 \right) + 3(c - R)\left(x_n^2 - x_2^2 \right) \right\} \qquad (4.297)$$

$$C_s x_s = V_{2b} x_{2b} + V_{1c} x_{1c} + V_{1d} x_{1d} - V_{3c} x_{3c} \qquad (4.298)$$

$$T_s x_t = V_{3a} x_{3a} + V_{3b} x_{3b} \qquad (4.299)$$

Case 8: $c_B < c < c_{c1}$. V_{2b} = Equation 4.282, $V_{2b} x_{2b}$ = Equation 4.287, V_{1d} = Equation 4.281, $V_{1d} x_{1d}$ = Equation 4.286, V_{1c} = Equation 4.292, and c_{c1} = Equation 4.183.

$$V_{3b} = -(k_1/6)(k_4)\{(c - R)^2(c - 4R) - R^2(4R - 3c)\} \qquad (4.300)$$

$$V_{3c} = (k_1/6)(k_4)\left\{ x_2 \left[-2x_2^2 + 3x_2 c - 6R(c - R) \right] - (c - R)^2 (c - 4R) \right\} \qquad (4.301)$$

$$C_s = V_{2b} + V_{1c} + V_{1d} + V_{3c} \qquad (4.302)$$

$$T_s = V_{3b} \qquad (4.303)$$

$$V_{3b}x_{3b} = (k_1/12)(k_4)\{(c-R)^3(3R-c) - R^3(2c-3R)\} \quad (4.304)$$

$$V_{3c}x_{3c} = -(k_1/12)(k_4)\left\{x_2^2\left[3x_2^2 - 4x_2c + 6R(c-R)\right] - (c-R)^3(3R-c)\right\} \quad (4.305)$$

$$V_{1c}x_{1c} = (k_1/3)(k_3)\left\{2\left(x_n^3 - x_2^3\right) + 3(c-R)\left(x_n^2 + x_2^2\right)\right\} \quad (4.306)$$

$$C_s x_s = V_{2b}x_{2b} - V_{1c}x_{1c} + V_{1d}x_{1d} - V_{3c}x_{3c} \quad (4.307)$$

$$T_s x_t = V_{3b}x_{3b} \quad (4.308)$$

Case 9: $c_{c1} < c < 2R$. V_{2b} = Equation 4.282, $V_{2b}x_{2b}$ = Equation 4.287, V_{1d} = Equation 4.281, $V_{1d}x_{1d}$ = Equation 4.286, V_{3b} = Equation 4.300, $V_{3b}x_{3b}$ = Equation 4.304, V_{3c} = Equation 4.301, and $V_{3c}x_{3c}$ = Equation 4.305.

$$V_{1c} = (k_1 k_3)\left\{x_n^2 - x_2^2 - 2(c-R)(x_n - x_2)\right\} \quad (4.309)$$

$$C_s = V_{2b} + V_{1c} + V_{1d} + V_{3c} \quad (4.310)$$

$$T_s = V_{3b} \quad (4.311)$$

$$V_{1c}x_{1c} = -(k_1/3)(k_3)\left\{2\left(x_2^3 - x_n^3\right) + 3(c-R)\left(x_n^2 - x_2^2\right)\right\} \quad (4.312)$$

$$C_s x_s = V_{2b}x_{2b} - V_{1c}x_{1c} + V_{1d}x_{1d} - V_{3c}x_{3c} \quad (4.313)$$

$$T_s x_t = V_{3b}x_{3b} \quad (4.314)$$

Case 10: $2R < c < c_2$. V_{2b} = Equation 4.282, $V_{2b}x_{2b}$ = Equation 4.287, $V_{1d}x_{1d}$ = Equation 4.286, and

$$c_2 = [e_c/(e_c - e_s)](R + x_2) \quad (4.315)$$

$$V_{1d} = 2k_2 k_3 (x_2 + x_n) \quad (4.316)$$

$$V_{1c} = (k_1 k_3)\left\{x_n^2 - x_2^2 - 2(c-R)(x_n - x_2)\right\} \quad (4.317)$$

$$V_{3d} = (k_1/6)(k_4)\left\{x_2\left[-2x_2^2 + 3x_2c - 6R)(c-R)\right] - R^2(4R-3c)\right\} \quad (4.318)$$

$$C_s = V_{2b} + V_{1c} + V_{1d} + V_{3d} \quad (4.319)$$

$$T_s = 0 \quad (4.320)$$

$$V_{1c}x_{1c} = -(k_1/3)(k_3)\left\{2\left(x_2^3 - x_n^3\right) + 3(c-R)\left(x_n^2 - x_2^2\right)\right\} \tag{4.321}$$

$$V_{3d}x_{3d} = -(k_1/12)(k_4)\left\{x_2^2\left[3x_2^2 - 4x_2c + 6R(c-R)\right] - R^3(2c-3R)\right\} \tag{4.322}$$

$$C_s x_s = V_{2b}x_{2b} - V_{1c}x_{1c} + V_{1d}x_{1d} - V_{3d}x_{3d} \tag{4.323}$$

$$T_s x_t = 0 \tag{4.324}$$

Case 11: $c_2 < c < c_3$. V_{2b} = Equation 4.282, $V_{2b}x_{2b}$ = Equation 4.287, and

$$c_3 = [e_c/(e_c - e_s)]2R \tag{4.325}$$

$$V_{1d} = 4k_1k_3x_2 \tag{4.326}$$

$$V_{3d} = (k_2/2)(k_4)\left\{x_2^2 - x_n^2 - 2R_0(x_2 - x_n)\right\} \tag{4.327}$$

$$V_{3b} = (k_1/6)(k_4)\left\{x_n\left[-2x_n^2 + 3x_nc - 6R(c-R)\right] - R^2(4R-3c)\right\} \tag{4.328}$$

$$C_s = V_{2b} + V_{1d} + V_{3d} + V_{3b} \tag{4.329}$$

$$V_{1d}x_{1d} = 0 \tag{4.330}$$

$$V_{3d}x_{3d} = -(k_2/6)(k_4)\left\{2\left(x_n^3 - x_2^3\right) + 3R\left(x_2^2 - x_n^2\right)\right\} \tag{4.331}$$

$$V_{3b}x_{3b} = -(k_1/12)(k_4)\left\{x_n^2\left[3x_n^2 - 4x_nc + 6R(c-R)\right] - R^3(2c-3R)\right\} \tag{4.332}$$

$$C_s x_s = V_{2b}x_{2b} - V_{3b}x_{3b} + V_{1d}x_{1d} - V_{3d}x_{3d} \tag{4.333}$$

Case 12: $c > c_3$. V_{2b} = Equation 4.282, $V_{2b}x_{2b}$ = Equation 4.287, V_{1d} = Equation 4.326, and $V_{1d}x_{1d}$ = Equation 4.330.

$$V_{3d} = (k_2/2)(k_4)\{R^2 + x_2(x_2 - 2R)\} \tag{4.334}$$

$$C_s = V_{2b} + V_{1d} + V_{3d} \tag{4.335}$$

$$V_{3d}x_{3d} = -(k_2/6)(k_4)\left\{x_2^2(-2x_2 + 3R) - R^3\right\} \tag{4.336}$$

$$C_s x_s = V_{2b}x_{2b} + V_{1d}x_{1d} - V_{3d}x_{3d} \tag{4.337}$$

Equations 4.105 to 4.337 are the solution for the capacity of a rectangular CFT column consisting of the axial capacity, P, and bending moment capacity,

M, parallel to the capacity axis. These equations are programmed in Excel spreadsheets to obtain the numerical values that can be plotted to yield the capacity curve of this rectangular CFT column. The equations for the stress variables were written for the specific rectangular CFT column in the example when the capacity axis is along the diagonal.

For other positions of the capacity axis, the stress variables change, and therefore, the limits of the stress volumes have to be adjusted and the applications of the derived equations adjusted accordingly. Hence, it is important to draw the stress diagrams for each case of a rectangular CFT column, as this will show the proper limits of the stress–strain diagram to be evaluated since the dimensions of the rectangular CFT column also affect the limits of the uniform and triangular stress–strain diagram.

The capacity curve along the diagonal is sufficient to compare the external loads. Before the external bending moment can be compared to the calculated capacity, it is necessary to obtain the component of the resultant external bending moment along the capacity axis. Since the location of the resultant external load with reference to the horizontal axis is known, resolving the component of the result along the capacity axis is easy since the position of the capacity axis from the horizontal is also known.

The capacity of a square CFT column is obtained when $b = d$ in the preceding equations or the same numerical value can be substituted in the Excel spreadsheets for the variables b and d.

The resultant bending moment capacity of the rectangular CFT column can be obtained by writing the equations of the bending moment perpendicular to the capacity axis. For details of the equations to use see Section 4.4.

From the Excel spreadsheets, we can selectively isolate the values of the capacity at key points on the capacity curve. The numerical values at these points are convenient references for checking that the sets of external loads are within the boundaries of these key points.

We shall now provide examples of circular CFT columns and rectangular CFT columns and plot the capacity curves of these two basic sections in structural analysis. The shape of these capacity curves should follow a smooth curvature to ensure that Equations 4.1 to 4.89 (for circular CFT section) and Equations 4.90 to 4.337 (for rectangular CFT section) are accurate.

It is worth noting that in the above analysis we did not invoke the standard interaction formula for biaxial bending or the Euler column formula to obtain the capacity of a circular or a rectangular CFT column section.

We shall show that the standard interaction formula for biaxial bending is not an accurate method of predicting the capacities of a rectangular CFT column section.

4.3 Capacity curve

A capacity curve is the plot of the numerical values of the concrete and steel capacities from the Excel spreadsheets. The vertical axis of this capacity curve represents the axial capacity, and the horizontal axis represents the moment capacity of a concrete-filled steel tube.

Example 4.1

An 8 in. (203.2 mm) diameter steel pipe with a shell thickness of 0.319 in. (8.1 mm) is filled with a 5 ksi (34.5 MPa) concrete and steel yield stress of 50 ksi (344.8 MPa). Determine the capacity curve of this concrete-filled steel pipe.

> **Solution:** Program the above equations for the steel forces with Excel spreadsheets using the given values of a circular CFT column. Figure 4.6 is the steel capacity curve obtained from the Excel spreadsheets. For the concrete forces, use the derived Equations 3.4 to 3.17 for a circular concrete section in Chapter 3 and also program these equations in Excel spreadsheets. Figure 4.7 is the concrete capacity curve obtained from the spreadsheets. To obtain the capacity of the circular CFT column section, add the concrete and steel forces together. Figure 4.8 is the resulting capacity curve.

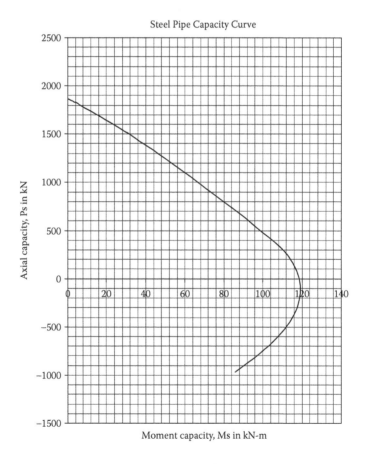

Figure 4.6 Steel capacity.

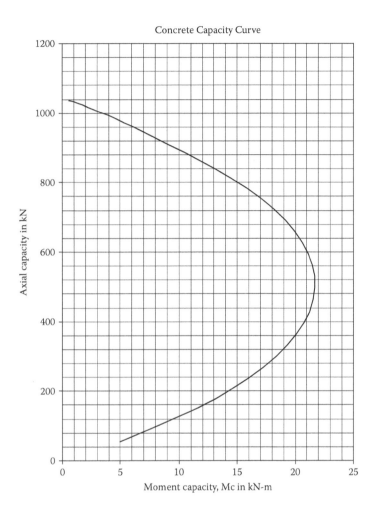

Figure 4.7 Concrete capacity.

The plot of the capacities of the circular CFT column section is a smooth curve, indicating that the equations used in the analysis are correct. This curve defines the relationship of the axial capacity with the moment capacity. When the moment capacity is increased, the axial capacity is decreased. This characteristic is common to the capacity curves for steel and concrete materials.

The steel capacity curve shows values in compression as well as tension. The tension part is below the zero line, indicating the tensile capacity of the steel pipe as it develops its internal capacity to resist any external loads. The concrete capacity curve can only utilize its compressive strength because in practice the tensile capacity of concrete is not reliable and should not be used to resist

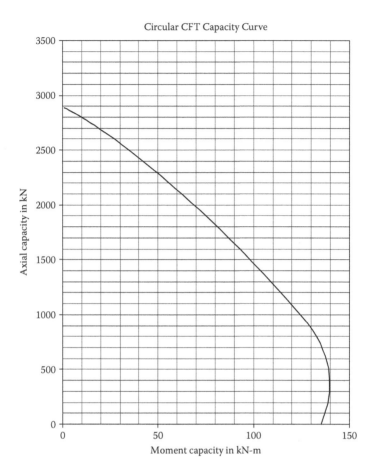

Figure 4.8 CFT steel pipe capacity.

external loads. Hence, only the parabolic compressive stress diagram is applied to develop the capacity of a concrete section.

The capacity curve intersects the vertical axis at a point where moment is zero in the section, and the point of intersection is the measure of the theoretical maximum axial load in the column. However, because of the inherent imperfection in the straightness of a column length, there will always be a certain amount of eccentricity of the applied axial load. Also, at points of support of a column, end moments are introduced, and thus the stress at these points consists of direct stress and bending. This is why most building codes require a minimum eccentricity in column designs. This additional bending moment should be considered as an added external load.

Example 4.2

Plot the capacity curve along the diagonal of a concrete-filled rectangular steel tubular column with the concrete core dimension b = 254 mm (10 in.) and d = 355.6 mm (14 in.), with shell thickness of 25.4 mm (1 in.), $f_c' = 32$ MPa (5 kips/in.²), and $f_y = 345$ MPa (50 kips/in.²).

> **Solution:** Figure 4.9 is the plot of the concrete capacity from Excel spreadsheets using Equations 3.18 to 3.45 in Chapter 3. Figure 4.10 is the plot of the rectangular steel tubing capacity from Excel spreadsheets using Equation 4.105 to Equation 4.337. Figure 4.11 is the capacity curve of the rectangular CFT column determined by adding the concrete capacity to the steel capacity.

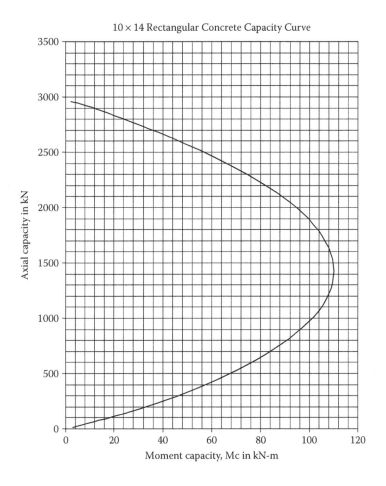

Figure 4.9 Concrete capacity.

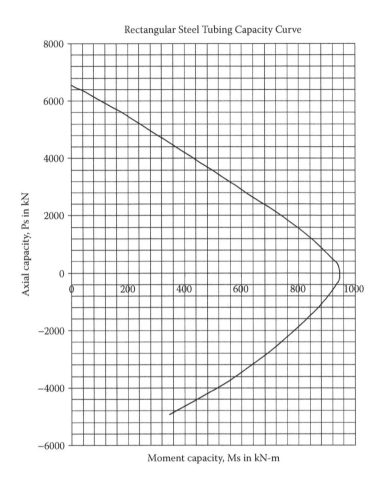

Rectangular Steel Tubing Capacity Curve

Figure 4.10 Steel capacity.

Once an example is solved using the Excel spreadsheets, it is just a matter of changing the numerical values of the variables to solve for any other rectangular or square CFT column sections. The structural engineer's task becomes easier because all he or she has to determine is the numerical values of external loads on a rectangular CFT column. When the external loads are known, all that is left to do is compare these external loads to the internal capacities of the column from the Excel spreadsheets.

4.3.1 Key points in the capacity curve

Figure 4.12 shows the schematic positions and numerical values at key points of the capacity curve of Example 4.1. The numerical values may be

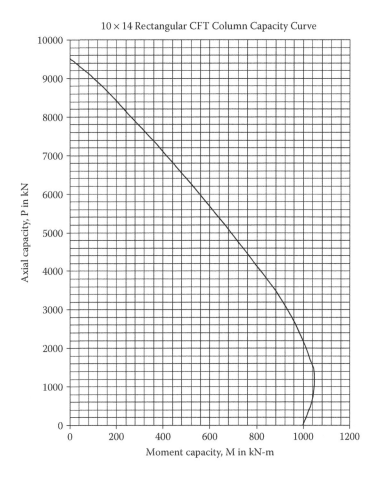

Figure 4.11 Rectangular CFT column capacity.

approximated from the derived equations by assigning values of the compressive depth c as follows:

- At key point S, $c = R + R_o$
- At key point Q, $c = 4 R_o$
- At key point T, $c = R$
- At key point U, $c = 0.85 R$

Although the results obtained may be approximate and not as exact as those from the Excel worksheets, they are good enough to check the capacity of a steel pipe filled with concrete.

$c = (R_o + R)$ when the tension in steel is zero. The value of P is the minimum recommended axial capacity to account for minimum load

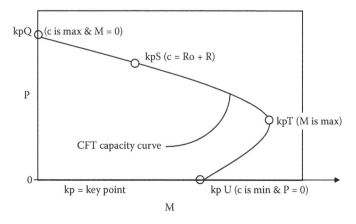

Key point	c, inches	M, in-kips	P, kips	c, cm	M, kN-m	P, kN
Q	17	0	651	43.2	0	2897
S	8.3	514	487	21.1	58	2166
T	4	1234	90	10.2	139	400
U	3.3	1213	0	8.4	137	0

Figure 4.12 Schematic locations of key points & values for a circular CFT column.

eccentricity. The theoretical maximum axial capacity when $M = 0$ and the beam moment capacity when $P = 0$ can be obtained from results shown in the Excel spreadsheets.

Figure 4.13 shows the positions and numerical values at key points of the capacity curve of Example 4.2. Values at these key points may be approximated using the derived equations by assigning values of the compressive depth c as follows:

- At key point S, $c = R + R_o$
- At key point Q, $c = 1.56 (R + R_o)$
- At key point T, $c = 0.48 (R + R_o)$
- At key point U, $c = 0.32 (R + R_o)$

Although the results may be approximate and not as exact as those from the Excel worksheets, they are good enough to check of the capacity of a rectangular steel tubing filled with concrete.

$c = (R_o + R)$ when the tension in steel is zero. The value of P is the minimum recommended axial capacity to account for minimum load eccentricity. The theoretical maximum axial capacity when $M = 0$ and the beam moment capacity when $P = 0$ can be obtained from results shown in the Excel spreadsheets.

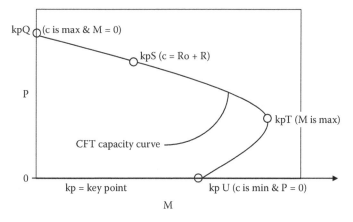

Key point	c, inches	M, in-kips	P, kips	c, cm	M, kN-m	P, kN
Q	29	0	9517	73.7	0	2140
S	18.6	302	7759	472.4	2673	1744
T	9	1047	1355	22.9	9270	305
U	6	997	0	15.28	636	0

Figure 4.13 Schematic locations of key points & values for a rectangular CFT column.

4.4 Column capacity axis

For a rectangular column section, the column capacity axis will establish the equilibrium of external and internal forces. For the equilibrium condition, the point of application of external forces must coincide with the centroid of the internal forces; that is, the external and internal forces must be collinear. Figure 3.13 in Chapter 3 shows how the external load has to be applied to a column section to ensure this equilibrium. The governing equations are repeated here for completeness of the analysis.

$$e = e_u \cos(\theta - \theta_u) \qquad (4.338)$$

$$P = P_u \qquad (4.339)$$

in which e = the eccentricity of the column capacity (M/P), e_u = the eccentricity of the loading condition (M_u/P_u), θ_u = the angle of inclination of the loading condition (M_2/M_1), and θ = the angle of the column capacity axis with the horizontal axis.

Equation 4.338 indicates that the resultant external load falls below the column capacity axis for a particular value of c in the rectangular column section. In other words, we cannot use the column capacity along the load axis since the resultant external load will not coincide with the center of gravity of internal forces of a column section defined by the value of c.

However, the author recommends that, for design purposes, we assume the diagonal as the reference axis for column capacity since the moment capacity increases toward the uniaxial position of the column capacity at $\theta = 0$.

This assumption enables the preparation of column capacity curves or tables for any given column section. It also ensures that the values being used are within the envelope of the column capacity curve. Once the plot of the capacity curve is obtained, the external load $P_u e_u$ should be resolved along the diagonal such that the external moment given by $P_u e_u \cos(\theta - \theta_u)$ can be plotted on the capacity curve. If the external moment is within the envelope of the capacity curve, the supplied section is adequate to support the external loads.

To check for equilibrium conditions for a column section subjected to an external biaxial load equal to $P_u e_u$, the following equations should be utilized:

$$\sum Vz = P_u e_u \sin(\theta - \theta_u) \tag{4.340}$$

$$\bar{z} = \sum Vz / \sum V = e_u \sin(\theta - \theta_u) \tag{4.341}$$

$$\sum V = P_u \tag{4.342}$$

These relationships involve trial and error. For our purposes, it is much simpler during the design analysis to plot the external loads on the envelope of the capacity curve. To ensure safety, the internal capacity should always be greater than the external load so that all external load combinations such as those in bridge design loading are inside the envelope of the column capacity curve.

To determine the value of the resultant bending moment, the moment around the capacity axis, M_z, is required. This moment is obtained using the following equations and the principle of superposition.

$$V_2 z_2 = 0.50(\cot\theta - \tan\theta)V_2\,\bar{x}_2 + [z_0 - 0.25\,h\,(\cot\theta - \tan\theta)]V_2 \tag{4.343}$$

$$V_1 z_1 = \tan\theta\,(V_1\,\bar{x}_1) \quad \text{when} \quad \theta < [(\pi/2) - \alpha] \tag{4.344}$$

$$V_1 z_1 = -\cot\theta\,(V_1\,\bar{x}_1) \quad \text{when} \quad \theta > [(\pi/2) - \alpha] \tag{4.345}$$

$$V_3 z_3 = -0.50(\cot\theta - \tan\theta)\,V_3\,\bar{x}_3 + [0.25\,h(\cot\theta - \tan\theta) - z_0]V_3 \tag{4.346}$$

Equations 4.343 to 4.346 are applied to the outer rectangular area and to the inner rectangular area to obtain the value of M_z. The resultant moment capacity is then obtained by the expression

$$M_R = \left\{ M_z^2 + M^2 \right\}^{1/2} \tag{4.347}$$

4.4.1 Variation of moment capacity

As the column capacity axis is varied from zero to $\pi/2$, the moment capacity also varies accordingly. Figure 4.14 shows the capacity curve when θ is zero, and Figure 4.15 is the capacity curve for Example 4.2 when θ is $\pi/2$.

It is obvious from this demonstration of moment capacity that the standard column interaction formula for biaxial bending, which gives only one set of column capacity values, is no longer warranted in structural analysis. It is also clear that for biaxial bending conditions, development of the uniaxial capacity at $\theta = 0$ and $\theta = \pi/2$ is not possible. This alone makes the application of the standard interaction formula for biaxial bending not only incorrect but also a violation of the basic principle of structural mechanics that requires consideration of the geometric properties of the section.

4.4.2 Limitations of the standard interaction formula

The severe limitations of the standard interaction formula can be shown from the results of the analytical method. The sum of the ratios of the external

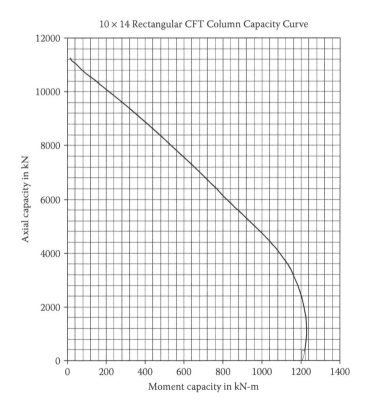

Figure 4.14 Capacity curve when $\theta = 0$.

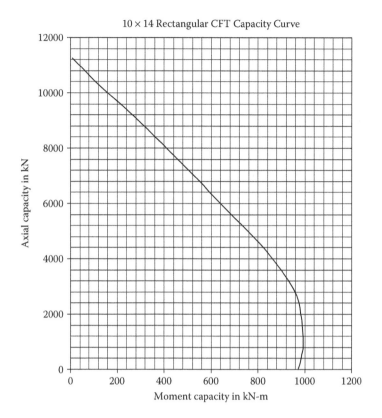

Figure 4.15 Capacity curve when $\theta = \pi/2$.

load to the internal capacity of a section in orthogonal axes 2-2 and 1-1 is added up to test for unity as a maximum value. From the tabulated values, it can be easily concluded that the application of the standard interaction formula will result in the underutilization of a structural member.

Table 4.1 shows the application of the interaction formula using the results of the analytical method for Example 4.2. Assume that the resultant internal moment capacity is equal to the external load. The given values at key point U from the Excel spreadsheets are $M_R = 1047$ kN-m, $P = 1268$ kN, $M_1 = 1229$ kN-m, $M_2 = 993$ kN-m, $P_{max} = 9485$ kN, and $\theta = 0.62025$ radian.

The result is resolved around the 1 and 2 axes as

$$M_1' = M_R' \cos \theta_u \quad \text{and} \quad M_2' = M_R' \sin \theta_u \qquad (4.348)$$

The ratios of external to internal moments are $R_1 = M_1'/M_1$ and $R_2 = M_2'/M_2$. $R = M_R'/M_R$ is the fraction of the resultant bending moment captured by using the standard interaction formula. Comparison between the analytical method and the standard interaction formula for biaxial bending can be made by using the results obtained above. The standard interaction formula for biaxial bending uses the sum of the ratios of the moments at orthogonal axes, which is not to exceed unity or one.

Table 4.1 is constructed such that $M_1' = M_R' \cos\theta$, $M_2' = M_R' \sin\theta$, $R_1 = M_1'/M_1$, $R_2 = M_2'/M_2$, and $R = M_R'/M_R$. Note the $\theta = \theta_u$ in this example for the equilibrium of external and internal forces.

From these tabulations, the standard interaction formula for biaxial bending can only capture 64% of the potential capacity of the rectangular CFT column section.

Notations

For circular CFT columns

c: compressive depth of a concrete core
c_B: concrete compressive depth when ultimate concrete and steel stresses are developed
e_c: concrete strain = 0.003 (United States) and = 0.0035 (Canada)
e_s: steel strain
t: wall thickness of a steel pipe
R: radius of a concrete core or the inner radius of a steel pipe
R_o: outer radius of a steel pipe
f_c': specified ultimate compressive strength of concrete
f_y: specified yield stress of a steel bar
M: ultimate moment capacity
M_c: moment capacity of concrete
M_s: moment capacity of a steel tube
P: ultimate axial capacity
P_c: axial capacity of concrete
P_s: moment capacity of a steel tube

Subscripts

c: concrete
s: steel tube

Note: All other alphabets or symbols are defined in the context of their use.

For rectangular CFT columns

b: width of a rectangular concrete section
c: depth of a concrete section in compression
d: depth of a rectangular concrete section
d': concrete cover from the edge to center of any steel bar
d_1: distance from the concrete edge to the first steel bar

e: eccentricity $= M/P$

e_c: useable concrete strain

e_s: steel yield strain

e_n: compressive (or tensile) steel strain at the nth bar location

e_u: eccentricity of the external load $(= M_u/P_u)$

f_c: compressive stress in concrete

f_c': specified ultimate compressive strength of concrete

f_y: specified yield stress of a steel bar

f_{yn}: steel stress $\leq f_y$

h: overall thickness of a member

x_m: location of a steel bar from a reference axis

x_n: = location of a steel bar from the concrete compressive edge

$\bar{z}$: distance from the X-axis of the centroid of internal forces

M_1: external moment around axis 1-1

M_2: external moment around axis 2-2

M: internal ultimate moment capacity

M_l: number of bars along the depth of a section

M_u: external resultant moment

N: number of bars along the width of a section

P: internal ultimate axial capacity

P_u: external axial load

R: radius of a circular column

θ: inclination of the column capacity axis with the horizontal axis

θ_u: arctan M_2/M_1 (inclination of the resultant external forces around the horizontal axis)

α_1: central angle subtended by one bar spacing in a circular column

α: arctan (b/d) = inclination of the diagonal of a rectangular column with the horizontal axis

ACI: American Concrete Institute

CRSI: Concrete Reinforcing Steel Institute

Note: All other alphabets and symbols used in mathematical derivations are defined in the context of their use.

References

American Institute of Steel Construction, *Manual of Steel Construction*, 7th ed., New York, 1977, 1.101.

Bowles, J. E., *Physical and Geotechnical Properties of Soils*, McGraw-Hill, 1979, 286.

Holtz, R. D. and Kovacs, W. D., *An Introduction to Geotechnical Engineering*, Prentice Hall, Englewoods Cliffs, NJ, 1981, 346.

Jarquio, R. V., *Analytical Method in Reinforced Concrete*, Universal Publishers, Boca Raton, FL, 2004.

Jarquio, R. V., Analytical method in structural analysis, in *Proceedings of the International Conference on Computing in Civil Engineering*, ASCE, 2005, 40794.

Jarquio, R. V., Limitations of the standard interaction formula for biaxial bending, *Hong Kong Institute of Engineers Transactions*, 12(2), 41, 2005.

Jarquio, R. V., Yield capacity curve of rectangular steel tubular section, in *Proceedings of Collaboration and Harmonization in Creative Systems*, Hara, T. Ed., Taylor and Francis, London, 2005, 615.

Jarquio, R. V., Limitations of the standard interaction formula for biaxial bending as applied to rectangular steel tubular columns, in *Proceedings of the Joint International Conference on Computing and Decision Making in Civil and Building Engineering*, Rivard, H., Miresco, E., and Melham, H., Eds, Montreal, 2006, 504.

Poulous, H. G. and Davis, E. H., *Elastic Solutions for Soil and Rock Mechanics*, John Wiley & Sons, New York, 1974, p. 40.

Smith, P. F., Longley, W. R., and Granville, W. A., *Elements of the Differential and Integral Calculus*, rev. ed., Ginn and Company, Boston, 1941, 286.

Steinbrenner, W., A rational method for the determination of the vertical normal stresses under foundations, in *Proceedings of the First International Conference on Soil Mechanics and Foundation Engineering, Section E: Stress Distribution in Soils*, Berlin, 1936, 142.

Terzaghi, K., *Theoretical Soil Mechanics*, John Wiley and Sons, New York, 1943, 373.

Index